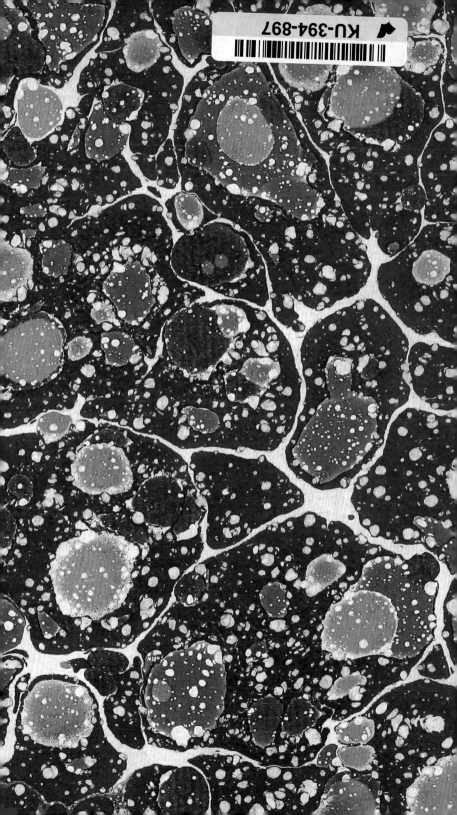

Further praise for *The Garden Against Time*

'A book that begins as beguiling and beautiful then flicks into the revelatory: the work of salvaging a ruined garden in Suffolk becomes a book about a different kind of salvation altogether. Her mind is so agile, so capacious, so widely ranging, so consistently surprising'
Neel Mukherjee

'Olivia Laing is a marvellous writer. So prepare yourself to be enchanted'
Jilly Cooper

'A magisterial work, and the exacting sensuality of her garden writing is pure pleasure, delight, surprise. A triumph, from a writer at the height of her powers'
Francesca Segal

'An inspiring and deeply thoughtful book – I loved it!'
Alison Light

'Powerful, reflective and captivating to read'
Fergus Garrett

'Olivia Laing has written a book about making her garden, which is by turns lyrical, consoling, disturbing and inspiring. It's a book for thinking gardeners everywhere'
Mary Keen

The Garden Against Time

ALSO BY OLIVIA LAING

To the River

The Trip to Echo Spring

The Lonely City

Crudo

Funny Weather

Everybody

OLIVIA LAING

The Garden Against Time

IN SEARCH OF

A COMMON PARADISE

PICADOR

First published 2024 by Picador
an imprint of Pan Macmillan
The Smithson, 6 Briset Street, London ECIM 5NR
EU representative: Macmillan Publishers Ireland Ltd, 1st Floor,
The Liffey Trust Centre, 117–126 Sheriff Street Upper,
Dublin 1, DOI YC43
Associated companies throughout the world
www.panmacmillan.com

ISBN 978-1-5-2906-6678

1 3 5 7 9 8 6 4 2

A CIP catalogue record for this book is available from the British Library.

Typeset by Palimpsest Book Production Ltd, Falkirk, Stirlingshire
Printed and bound by CPI Group (UK) Ltd, Croydon, CRO 4YY

Visit www.picador.com to read more about all our books
and to buy them. You will also find features, author interviews and
news of any author events, and you can sign up for e-newsletters
so that you're always first to hear about our new releases.

To the gardeners

and in memory of
Pauline Craig

And as it works, th' industrious bee
Computes its time as well as we.
How could such sweet and wholesome hours
Be reckon'd but with herbs and flow'rs!

Andrew Marvell, 'The Garden'

This boke is cald garthen closed, wel enseled,
paradyse ful of all appils.

Richard Rolle, *English Psalter*

CONTENTS

I   A Door in the Wall   1

II   Paradise   19

III   A Landscape Without People   59

IV   The Sovran Planter   103

V   Garden State   143

VI   Benton Apollo   185

VII   The World My Wilderness   225

VIII   The Expelling Angel   267

Notes   293

Bibliography   305

Acknowledgements   313

List of Illustrations   317

The Garden Against Time

I

A DOOR IN THE WALL

I have a dream sometimes, not often. I dream that I am in a house, and discover a door I didn't know was there. It opens into an unexpected garden, and for a weightless moment I find myself inhabiting new territory, flush with potential. Maybe there are steps down to a pond, or a statue surrounded by fallen leaves. It is never tidy, always beguilingly overgrown, with the corresponding sense of hidden riches. What might grow here, what rare peonies, irises, roses will I find? I wake with the sense that a too-tight joint has loosened, and that everything runs fluent with new life.

For most of the years that I have had this dream, I didn't have a garden of my own. I came to home ownership late, renting until I was forty, and only rarely in flats with outdoor space. The first of these temporary gardens was in Brighton. It was so narrow I could almost touch both fences at once, dropping away over the crest of the Downs in three precipitous terraces, culminating in a greenhouse with a rampant grapevine, inhabited by a golden-eyed toad.

I planted calendula there, pot marigold, which according to

the sixteenth-century herbalist Gerard would 'strengthen and comfort the heart very much'. I was training to be a herbalist and my head was full of plants, an entanglement of natural forms. The study of botany was an education in looking. It made the ordinary world more intricate and finely detailed, as if I had acquired a magnifying glass that trebled the eye's capacity. Each plant was so interwoven into human history that to study it was to tumble down a conduit through time. 'The Wild Marigold is like unto the single garden Marigold, but altogether lesser; and the whole plant perisheth at the first approach of Winter, and recovereth itself again by falling of the seed.'

In Cambridge a decade later I planted salvias and broom, and remade the stinking pond, which in the spring was full of newts, swimming to the surface to exhale a silver ball of air. I lived on short-term contracts, with black mould on the walls, but the gardens made me feel permanent, or maybe instead brought me to terms with transience. What I loved, aside from the work of making, was the self-forgetfulness of the labour, the immersion in a kind of trance of attention that was as unlike daily thinking as dream logic is to waking. Time stopped, or rather swept me up with it. In my twenties, I'd read a list of rules for being, and was so impressed that I copied them into my little black notebook, which in those days was full of aphorisms and advice about how to be a person. The rule I liked best stated that it is always worth making a garden, no matter how temporary your stay. Perhaps they wouldn't last, but wasn't it better to go on like Johnny Appleseed, leaving draughts of pollen in your wake?

Each of these gardens was a way of making myself at home, but all the same I longed for a permanent space of my own, more each time a landlord ended a contract and sold off the space that so evidently wasn't mine. I'd wanted it since childhood, even more painfully than I wanted a house. Love aside, it was the most consistent and consuming of all my desires, and as it happened the one thing brought the other into my life, a surge of good fortune I still can't quite believe. In my forties I met, fell in love with and soon married a Cambridge don, an inordinately clever, shy and affectionate man. Ian was much older than me and lived in a terraced house that was filled from floor to ceiling with books. His wife had recently died and just after I moved in he had two serious operations. We'd become friends in the first place because of our shared interest in gardening, and after he retired we began to talk about moving somewhere with the potential to restore a garden, or make one from scratch. It wasn't clear how long we'd have together, and creating a garden felt like the right way to spend at least some of that time.

During this period of searching, my aunt emailed me a photograph of a house enveloped to the gutter in roses, which had been trained in loose curves, so that sprays of flowers tapped against the windows. There were box squares on either side of the front door, clipped into the comical form of Mr Kipling's French Fancies. It looked exactly like the solid, four-square, chimneyed houses I'd drawn as a child, an embodiment of the rootedness I'd craved so badly in those rickety and uncertain years. I skipped through the descriptions of the interior until I got to the section titled 'Outside'. 'The RHS gardens are a particular feature of the house, laid out by the distinguished gardener Mark Rumary of Notcutts.' This was more promising! Though I hadn't heard of Mark Rumary, I knew of Notcutts, the famous Suffolk nursery that often won medals at Chelsea for its displays.

We went to look at it in January 2020, driving through little Suffolk villages until we were almost at the coast. With each mile the land grew flatter and the sky seemed to accept more light. We were so early I had time to eat poached eggs and toast in the cafe opposite, eyes trained on the clock. You couldn't see the garden from the street. It must be hidden round the back. I saw it as soon as the front door opened. The long dim hall ran down to a second door made of glass. A wave of verdurous light flooded in.

Outside, the trees were bare. The garden was walled, its soft red Suffolk brick covered in knots of climbers of many kinds: wisteria, clematis, winter jasmine and honeysuckle, alongside ramparts and streamers of ivy. Everything was neglected and overgrown, but even at a glance I could see unusual plants like witch hazel, its lemon-peel flowers exuding a hypnotic, astringent scent, and the unmistakable black buds of a tree peony. At the far end a door in the wall led into a Victorian coach house, now used as a makeshift garage. Beyond it was a loose box with an iron manger, exactly like the one in *The Children of Green Knowe*, where Tolly leaves sugar lumps for the ghost horse Feste. In the potting shed, the owner showed me Mark Rumary's cobwebbed gardening apron, still hanging from its hook.

The entire plot was just under a third of an acre, but it felt much larger because it had been so cunningly divided with hedges, one of beech and one of yew, so that you could never see the entirety at once, but continually passed through doorways and arches into secretive new spaces. One had a raised pond in the shape of a quatrefoil, and another seemed wholly abandoned, with rotting fruit trees, among them a medlar, a tree I only knew from Shakespeare's joke in *Romeo and Juliet* about what maids call the fruit: *open-arses*. There had been a wedding in there, and the ground was still covered in a circle of tarp, pierced by nettles and foxgloves. Beyond the far wall a slope of parkland encircled a rose-pink Georgian mansion, just visible through the bare branches of the sycamores. There was a curved door in this wall too, padlocked and painted a peeling duck-egg blue. Its presence had given rise to a rumour that this was once the dower house, though to me it echoed the enigmatic door in my garden dreams.

Many of the walls were latticed with roses. They looked as if they hadn't been pruned in years, and I thought, of course, of cross Mary Lennox with her jaundice-yellow skin, who pried her way into a garden like this and emerged a different sort of girl altogether. If I scraped these roses with a penknife I'd no doubt they'd be wick and green. Gardens have a knack for looking dead but rarely are, and anyway the ground was covered with snowdrops, pushing through rotting leaves. And then in the corner I spotted a daphne, larger than any I'd ever seen, its shell-pink clusters exhaling unsteady streams of sweetness. It was the first plant I'd fallen in love with, the first botanical name I'd learned as a child. More than anything, I wanted this garden to be mine.

That was January. Then it was February, and the first cases of Covid were being reported in the UK, followed by lockdowns in Italy, where the hospitals were already overwhelmed. At home, the prime minister, who would soon almost die of the virus, talked gaily about turning the tide. People began to wear masks, then to stay at home, then to worry over contamination by way of their post or shopping. Just after the spring equinox, a national lockdown was declared. Nearly everyone in the country was confined to their homes, permitted a single hour of outside exercise a day.

And so the world, which had lately moved so fast, simply stopped on its heels. In *Paradise Lost*, Milton describes the Earth hanging suspended from a chain. This is how it appears to Satan, as he journeys from Hell through the wastes of Chaos. First he sees Heaven, with its battlements of sapphire and opal, and then, 'fast by hanging in a golden Chain / This pendant world, in bigness as a Starr / of smallest Magnitude close by the Moon'. A small world, dangling: that's what the first suspended season of lockdown was like.

The weather was balmy, tender, almost foolishly lovely. As everything else contracted, spring brought a counter-surge of beauty, a non-stop froth of cherry blossom and cow parsley. Cambridge emptied of tourists and students. Even the playgrounds were shut, the swings padlocked to their supports.

Ian was in his seventies, with two aneurysms, and my panic over keeping him safe was already surging out of control. We walked through empty streets to empty parks, shying away when strangers did appear, holding our breath as joggers huffed past. Not that I went out often. A few weeks before lockdown, I'd developed a cough I couldn't shake, which turned into pleurisy. Feverish and bed-bound, I spent the long hours journeying in my mind to the garden, trying to find out everything I could about its genesis and evolution.

The current owners had lent me two essays about it, one by Rumary in an anthology of Suffolk gardens, and another from a 1974 issue of *Country Life*, illustrated with lavish black and white photographs. The story was written by the landscape architect Lanning Roper, who I later found out was Rumary's first employer. Delving around, I found another in *The Englishman's Garden*, edited by Rosemary Verey and Alvilde Lees-Milne. Rumary's tone was so warm and vivid it was as if he was standing in the room waving his hands, enthusiastic and self-deprecating, knowledgeably infatuated with plants. 'I'll never forget my excitement', he wrote, 'when I first saw this garden.' He'd moved to the house in 1961 with his partner, the composer Derek Melville, who even in 2000 he was still describing as his *friend*. Gay and closeted, a language I knew intimately from my own childhood in a gay family in the 1980s.

When they first arrived the garden was unkempt, 'with a small orchard of decrepit apples and plums growing in a luxuriant carpet of ground elder surrounded by unusually high walls which imparted the claustrophobic feeling of a prison yard.'

There was a network of narrow, crumbling paths that didn't seem to lead anywhere. The soil was light and sandy, and a belt of elms beyond the far wall cast a dense bolt of shade. Rumary had designed hundreds of gardens, but this was the only one he ever made for himself. Even at such a long remove his excitement was palpable. He dispensed with everything save a few mature trees, among them three Irish yews and a magnificent mulberry, planted in the reign of James I. Once the ground elder had been dealt with and the sick apple trees grubbed up, he realised the irregularity of the plot lent itself to being laid out in rooms that were like extensions of the house, using hedges to demarcate the boundaries; the classic Arts and Crafts style pioneered by Gertrude Jekyll and deployed so skilfully by Vita Sackville-West and Harold Nicolson at Sissinghurst Castle. In fact, the fig I'd admired on my visit derived from a cutting from one of the Sissinghurst trees.

In his original iteration of the garden, what was now the pond garden was planted with old-fashioned roses like 'Ferdinand Pichard' and 'Fantin Latour'. The space behind the yew hedge where the wedding marquee had been was a shady white garden, with different fruit trees illuminating each corner, including the Tibetan cherry and the winter-flowering cherry. Beneath these high cumuli of blossom were banks of hosta and bamboo, interplanted with a mass of skimmia, white potentilla and white phlox, along with white narcissus, regal lilies and lily flowered tulips. It was a good place to recover from a hangover, Rumary said, with a circular lawn like a green pool, the white flowers glowing in the narcotic light that filtered through the leaves. I imagined it as not so dissimilar to that creamy vision of summer evenings, *Carnation, Lily, Lily, Rose* by John Singer Sargent.

Rumary's elegant sense of structure was perpetually imperilled by his desire to try out new cultivars and species. 'Over the years there has been a constant battle between designer and plantsman,' he wrote, 'for while by training I belong to the former, by instinct I am of the latter. It is a Dr Jekyll and Mr Hyde situation, or in this case Miss Jekyll and Mr Hyde! In the early stages Jekyll had the upper hand . . . but Hyde would keep popping up and he still does.' I understood that kind of plant mania, being far more of a Hyde myself. The garden back then had been crammed with unusual and covetable plants. With the help of those three articles and two dozen pictures I'd rooted up online, I made a painstaking list of nearly two hundred of them, many chosen for their fragrance. I loved reading through it, distracting myself from the horrifying indeterminacy of the future by day-dreaming about the differing scents of Christmas box and wintersweet and *Rosa rugosa* 'Roseraie de l'Haÿ'. Some, I was sure, would have died or been removed over the decades. What had happened to the pineapple broom, which needs a warm south-facing wall to thrive? Was there still a spotted laurel, grown from a cutting taken at Chopin's grave, or pinks from seed gathered in George Sand's garden at Nohant?

As the terrors of the plague year grew, this half-imaginary, half-real garden became a place of solace to me, a zone apart that I could enter at will, even though I'd only seen it once.

This sounds like an idiosyncratic activity, but I was by no means alone in finding the garden a place of consolation that spring. While I was in bed, an unlikely obsession with gardening had taken hold. All across the world, people were engaged in a feverish new love affair with plants. Each morning my Instagram filled up with a sap-green tide of sweet pea seedlings and clematis spotted on daily walks. Over the course of 2020, three million people in Britain began to garden for the first time, over half of them under forty-five. Garden centres ran out of compost, seeds and even plants as people poured their energy into transforming the spaces in which they were confined. The same pattern was repeated globally, from Italy to India. In America, 18.3 million people started gardening in the pandemic, many of them millennials. The American seed company W. Atlee Burpee reported more sales in the first March of lockdown than at any other time in its 144 year history, while the Russian retailer Ozon reported a 30 per cent increase in seed sales. It was as if, during that becalmed and frightening season, plants had emerged into collective visibility, a source of succour and support.

It wasn't hard to understand why. Growing food is an instinct in times of insecurity, peaking during pandemics and wars. Gardening was grounding, soothing, useful, beautifying. It occupied the baggy days and provided a purpose for people abruptly untethered from office routines. It was a way of surrendering to the present moment in which we were all trapped. In a time interrupted – crouched on the threshold of unimaginable disaster, death toll soaring, no cure in sight – it was reassuring to see the evidence of time proceeding as it was meant to, seeds unfurling, buds breaking, daffodils pushing through the soil; a covenant of how the world should be and might again. Planting was a way of investing in a better future.

For some people, anyway. But the lockdown also made it painfully apparent that the garden, that supposed sanctuary from the world, was inescapably political. There was a grim disparity that sublime spring between the people pottering with trowels or typing from their deckchairs, and those trapped in tower blocks or mildewed bedsits. This disparity was only intensified as public parks and wild spaces were closed or subject to heightened policing, making them more inaccessible to the people who needed them most. According to research carried out by the Office of National Statistics that spring, the vast majority – 88 per cent – of the British population has access to a garden of some kind, including balconies, patios and communal garden spaces, but this distribution is by no means random. Black people are nearly four times as likely to have no access to a garden as white people, while people in unskilled and semi-skilled jobs, casual workers and the unemployed are almost three times as likely to be without a garden as those in professional or managerial positions. A study by the National Institute of Health in 2021 suggests that while gardens in America are less widely distributed among the population at large, white people are nearly two times as likely as Black or Asian citizens to have access to a garden.

As Black Lives Matter protests took hold around the world, gardens, and especially the aristocratic stately home gardens owned by the National Trust, were subject to scrutiny in their own right. A garden, a parkland might look more innocent, even virtuous, than a statue of a slave-trader, but they too have a hidden relationship with colonialism and slavery. It isn't just that many of our familiar garden plants, from yucca and magnolia to wisteria and agapanthus, are imported 'exotics', a legacy of a colonial-era mania for plant-hunting. Slavery also provided the capital for a concerted beautification of the landscape, as the grotesque profits from sugar plantations were used to found lavish houses and gardens back in England.

To certain audiences, this discussion was intolerable, politicising what they believed should be neutral, a beautiful haven from debate. They didn't want to question the cost of building paradise, or to have the cosy charm of the so-called heritage landscape undermined. To others, it made the garden a tarnished, even contaminated zone, a source of unquestioned privilege, the gleaming fruit of dirty money. Personally, I thought that while the spell of the garden does lie in its suspension, its seeming separation, from the larger world, the idea that it exists outside of history or politics is not a possibility. A garden is a time capsule, as well as a portal out of time.

The fact that owning a garden is a luxury, that access to land itself is a luxury, a privilege and not the right it should be, is hardly a new phenomenon. The story of the garden has from its Edenic beginning always also been a story about what or who is excluded or cast out, from types of plants to types of people. As Toni Morrison once observed: 'All paradises, all utopias are designed by who is not there, by the people who are not allowed in.' If some of England's seemingly sublime gardens were economically dependent on the sugar, cotton and tobacco plantations of America and the West Indies, others were contingent upon the practice of parliamentary enclosure, the legal process of taking the formerly open fields, commons and waste land of the medieval period into private ownership. Between 1760 and 1845, thousands of Enclosure Acts were passed. By 1914, over a fifth of the total area of England had been enclosed, a prelude to today's enraging statistic that half of the country is owned by less than 1 per cent of the population. Enclosure helped to facilitate a new arcadia: the great house in splendid isolation, islanded amidst an apparently natural parkland, which had been fastidiously manicured, stripped of its coarse human elements, from roads, churches and farmhouses to entire villages.

I'd been thinking about these more troubling aspects of the garden for a long time. By both income and inclination I'd spent far more of my life involved with ad-hoc gardens, established for very little on abandoned or degraded ground. I'd started training as a herbalist after a period of environmental activism, living for the first winter of my studies in a bender on an abandoned pig farm outside Brighton, part of a collective that was trying to make a community garden there. My decision to study herbal medicine stemmed from many seductive readings of *Modern Nature*, the filmmaker Derek Jarman's account of making a garden on the shingle beach at Dungeness, while he was dying of Aids. 'The Middle Ages have formed the paradise of my imagination,' he wrote, and his diary entries were interspersed with extracts from medieval herbals, a pharma-copoeia of magical and medicinal plants; rosemary, borage, heartsease, thyme. Each trailed so many memories and associations that his garden became two seemingly contradictory things at once: an escape hatch into the eternal and a way of stitching himself – though the word he actually uses is 'chained' – into the living landscape.

Around the time that the first cases of Covid were appearing in the news, I was involved in the campaign to save Derek Jarman's house, Prospect Cottage, and the famous garden that surrounded it. Two weeks into lockdown, the campaign reached what had felt like the impossibly ambitious crowdfunded target of £3.5 million. It seemed I wasn't the only one to find that improbable place sustaining, long after Jarman's death. His garden has no walls or fences, deliberately obliterating the border between cultivated and wild, the roses and red-hot pokers giving way to wind-sculpted clumps of sea kale and gorse. In this way it makes visible one of the most interesting aspects of gardens: that they exist on the threshold between artifice and nature, conscious decision and wild happenstance. Even the most mani-cured of plots are subject to an unceasing barrage of outside forces, from weather, insect activity and soil microorganisms to pollination patterns. A garden is a balancing act, which can take the form of collaboration or outright war. This tension between the world as it is and the world as humans desire it to be is at the heart of the climate crisis, and as such the garden can be a place of rehearsal too, of experimenting with inhabiting this relation-ship in new and perhaps less harmful ways.

As I knew from my own experiences, the story of the garden does not always enact larger patterns of privilege and exclusion. It's also a place of rebel outposts and dreams of a communal paradise, like that of the Diggers, the breakaway sect of the English Civil War, who made the still-radical assertion that the earth is a 'common treasury', imagine, for everyone to share. They established their brief vegetal Eden on St George's Hill in Surrey, in what is now a gated community favoured by Russian oligarchs. This kind of garden is a place of possibility, where new modes of living and models of power can and have been attempted, a container for ideas as well as a metaphor by which they can be expressed. As the artist Ian Hamilton-Finlay, who built the sculpture garden Little Sparta in Scotland, once observed, 'some gardens look like retreats but they are actually an attack.'

If I did get Rumary's garden, I would restore it, I told myself, but I would also trace how it had intersected with history, as even the smallest garden invariably must, since every plant is a traveller in space and time. I wanted to explore both types of garden stories: to count the cost of building paradise, but also to peer into the past and see if I could find versions of Eden that weren't founded on exclusion and exploitation, that might harbour ideas that could be vital in the difficult years ahead. Both of these questions felt very urgent to me. We were poised on the hinge of history, living in the era of mass extinction, the catastrophic endgame of humanity's relationship with the natural world. The garden could be a refuge from that, a place of change, but it can and has also embodied the power structures and mindsets that have driven this devastation.

There was something else, too. I was exhausted by the perpetual, agonised now of the news. I didn't just want to journey backwards through the centuries. I wanted to move into a different understanding of time: the kind of time that moves in spirals or cycles, pulsing between rot and fertility, light and darkness. I had an inkling even then that the gardener is initiated into a different understanding of time, which might also have a bearing on how to preclude the apocalypse we seem bent on careering into. I wanted to dig down, and see what I could find. A garden contains secrets, we all know that, buried elements that might put on strange growth or germinate in unexpected places. The garden that I chose had walls, but like every garden it was interconnected, wide open to the world.

II

PARADISE

We finally moved in mid-August. Dog days of summer, the ground parched and cracked. It was thirty-one degrees by twelve o'clock, the air wobbling like jelly. As soon as we got the keys, I went straight out into the garden. It wasn't how I remembered it, not at all. It looked neglected and clapped out, the lawn crisped, the borders sagging. The box had blight. The greenhouse was full of half-dead tomato plants, and outside its door the *Magnolia grandiflora* had dropped hundreds of leaves, the colour and consistency of baseball gloves. Had we made a mistake? Had I imagined it, the mood that seemed to gather between the walls? Later, I would learn that the garden was always at its worst in August, overexposed and broiling in the sun.

The movers arrived, and one by one the lovely empty rooms filled up with boxes. While they worked inside, I stripped the greenhouse, hauling ancient pots and bags of compost onto the grass, despite the heat. I scrubbed the staging and stowed my tools, stacking canes and plant supports in orderly piles. It was a bodge job, frankly. The building was in a far worse state than

I'd realised on that first seductive visit. Thick tendrils of ivy had worked under the roof, and the joists were so rotten you could press a thumb right through them. Many of the bricks in the back wall were blown, or stained a virulent algal green. Not a quick lick of paint, then: a full rebuild.

We hadn't lifted any of the plants from our old garden, not even the striped peony I'd been coaxing along since its infancy, but we did have a multitude of pots, including my burgeoning collection of nearly forty pelargoniums. Their names sounded like characters from a Jane Austen novel: 'Lady Plymouth', 'Lord Bute', 'Ashby', 'Brunswick', 'Mrs Stapleton'. As they came off the lorry I carried them to the pond garden one by one, arranging them around the sides of the raised pool along with a stone lion, head on paws, and two fistfuls of white pebbles Ian had once brought back from Greece. The water was thick with blanket weed. Banks of lady's mantle were foaming onto the flags, and in the far border a single cardoon was in full sail, crowns of imperial purple burning in the unsteady light. A mauve geranium ramped through scrawny roses. 'Rozanne'? It was so hot, the only sound the bees and the distant traffic on the A12.

After the movers left, we sat on the grass eating cherries. Long shadows were creeping up the lawn. At dawn that morning, my father had emailed to say his wife had deteriorated in the night. She'd been in the National Hospital for Neurology and Neuroscience since May, the latest in a long series of hospitalisations for a brain tumour. It was the first time in a decade he hadn't been able to visit her. On every previous occasion he'd travelled up to London each day by train to sit with her or speak to the staff, managing the complex orchestration of her care, which fell between oncology and neurology. When she was at home he looked after her alone, a situation that was becoming increasingly untenable. A few days ago we'd been discussing carers for her imminent discharge. But now she was gravely ill, and not expected to survive the week.

We moved on Monday. By Wednesday it seemed as if she was going to rally, pull off one of those startling volte-face recoveries she'd managed so many times before. Her blood pressure came back to normal, as did her body temperature. But on Friday morning, a few hours after this last cheerful bulletin, she died, and he rang weeping from Queen Square, where he was sitting on a bench in front of the hospital, holding a box of her clothes.

It was my father who'd instilled a love of gardening in me. My parents had divorced when I was four, and he'd spent his custody weekends taking us to every National Trust and stately home garden within a hundred miles of the M25, a circle we travelled continually since we lived at the bottom and he lived at the top. I was an anxious, not very happy child, and I loved the self-forgetfulness that happened in a garden. One of our favourite places to visit was Parham, an Elizabethan house in Sussex, where there were tiny formal pools in the corner of each of the walled gardens, covered with a glossy green baize of duckweed. Sissinghurst we went to in April, my birthday month, when the lanes were festive with apple blossom and you could get drunk on the ripe smell of wallflowers, baking in the sun under Vita's tower.

My father always carried a small black notebook on these expeditions. In it he made lists of plant names in his unreadable hand, pouncing on any unusual cultivar. I liked the old varieties in particular, hoarding away lists of antique roses or apples fashionable in sixteenth-century orchards. 'Winter Queening', 'Catshead', 'Golden Harvey', 'Green Custard', 'Old Permain'. His wife rarely accompanied us. In the box handed over by the hospital there was an A4 page on which she'd listed personal information, presumably assisted by a nurse, since she could no longer see to read or write. The final box was for things she disliked or preferred not to talk about. All it said was: 'Gardening – my husband normally does this.'

His desolation now, and the sense of shock and horror that radiated from her unexpected death, saturated my experience of the garden in those first days, magnifying the mood of apprehension generated by the pandemic. In the winter, I'd only seen the loveliness of the structure, the sense of promise. I hadn't really taken in how neglected it all was. Now I looked with different eyes. Trees bubbled with fungus, or grew in strange disordered forms, strung with luxuriant garlands of bindweed. Plants looked leggy or stunted, starved of nutrients, swamped by thugs that had long since slipped their bounds. One afternoon, a ribes collapsed right in front of me, and the tree above it seemed to be dying too. The previous owners had done an incredible job of restoring the house, but as they said themselves, they weren't gardeners, and during the months of lockdown they'd had no help, either.

It wasn't that I liked tidy gardens. I agreed with Frances

Hodgson Burnett's manifesto in *The Secret Garden*, that a garden loses all its charm if it becomes too spick and span. You have to be able to lose yourself, to feel, as Burnett put it, almost shut out of the world. One of the things I particularly liked about this garden was its odd sense of proportion: the way the walls and the beech hedge with its pair of arches were so absurdly high that it made us seem like Lilliputians passing through with our rakes. The height achieved by some of the plants seemed to emphasise the absorptive quality of even the most quotidian of garden tasks.

There's something magical about this removal from daily life, a vegetal transition that can feel almost ecstatic. At the same time, a garden that's too neglected can be uncanny, disturbing in the same way that an abandoned house is disturbing, the domestic invaded by agents of disorder and decay. The image of an untended garden circulates through literature as a metaphor for neglect on a larger scale. Think of the servant in *Richard II* who refuses to bind up the 'dangling apricocks' in the Duke of York's garden, asking why he should weed an enclosed plot when the garden of England is so unkempt, 'her fairest flowers chocked up, / Her fruit-trees all upturned, her hedges ruin'd, / Her knots disorder'd and her wholesome herbs / Swarming with caterpillars.' Horror lifts off those lines in waves. Something wrong, something rotten or infested that should be orderly and fruitful.

If the garden here stands squarely for the nation state, in *Hamlet*, written five years later, it conveys an emotional as well as political landscape. In the wake of his father's death, Hamlet compares the newly damaged and contaminated world to 'an unweeded garden /That grows to seed; things rank and gross in nature /Possess it merely.' The point of this persistent and unsettling image is that a garden is supposed to be planted and tended, and so its dissolution suggests a more sinister lapse than a patch of wilderness or waste ground would.

There's a similar species of anxiety in the eerie 'Time Passes' section of Virginia Woolf's *To the Lighthouse*, in which the Ramsays' house is left to the elements, also as a consequence of an unexpected death. Poppies seed themselves among the dahlias and artichokes engulf the roses, just as they had done in the pond beds here. There's a fecund, if horticulturally improbable, pleasure to this misrule, as the carnation mates with the cabbage. But there's also an acknowledgement that it can turn on an instant into something more desolate, even deathly. 'One feather, and the house, sinking, falling, would have turned and pitched downwards to the depths of darkness.' Then the house itself would shatter, its broken parts buried beneath a coverlet of hemlock and briars.

Those lines came back to me often in the early weeks. It was all so in need of care. *Nettles*, I wrote gloomily in my garden diary, *couch grass, brambles, alkanet*. No hemlock, but instead hogweed, taller than me, hundreds of white flower heads already gone to seed. The soil was like brown sugar, just as Mark Rumary had said. Pure Suffolk sand, which can't hold nutrients and needs regular enriching doses of organic matter. I hardly ever saw a worm, and each time I ventured into the ivy-covered regions at the back of a bed, I'd stumble on a corpse, the sinister stump of a tree or shrub, killed off by unknown causes and dressed in funereal weeds.

I spent the first weeks discovering the lay of the land, adjusting the map in my head to reality. It was four small gardens, really, plus a few stray beds beside the path that led along the north side of the house and past the potting shed, entering the main garden through another door in the wall, this one painted cream and sagging from its hinges. The house was on your left as you passed through it, a rosy L, higgledy piggledy from centuries of additions, one facet almost hidden by a wisteria, the fat coils of which prevented the sitting-room window from opening. Tucked inside the L was a paved terrace just big enough for a table, flanked by a little box parterre crammed with peonies and roses.

This terrace led onto another secretive garden on the south side of the house, unimaginatively called the greenhouse garden for its central and most alluring exhibit. There was a handkerchief of parched lawn with a plum tree at its centre, hemmed by hunched, unhappy shrubbery, including mahonias, arbutus and a magnolia I didn't recognise, as well as two apple trees, bafflingly planted in deep shade. The wall had been raised against the road by a screen of pleached hornbeams, tethered to municipal stakes. It looked like a hospital car park. I imagined sweet woodruff under there, rising in a green and white wash, a spring tide.

My diary was full of notes like this, worried accounts interspersed with dreams of fertility and repair. Frankly, I felt overwhelmed. Where was I supposed to start? What principle should I apply? I'd renovated gardens before, but never one with such an exalted legacy. I didn't want to ruin anything, to toss out pearls in ignorance or error. My friend Simon, the head gardener at Worcester College in Oxford, gave the best advice. He told me not to take out anything for a full year, unless I was absolutely certain of its identity and appearance right through the seasons. It wasn't in my nature to be patient, but if I wanted to know what had survived, the only fail-safe method was to watch and wait.

The main garden was opposite the house. It had been laid to lawn and edged with deep curving borders that culminated at the far end in the yew hedge that hid the wedding garden. The north border was dominated by a *Magnolia* x *soulangeana*, that once-familiar tree of suburban gardens, which produces a hyper-abundance of beak-shaped pink flowers so early that they often succumb to frost, covering pavements in brown slime. I used to regard the one outside my grandmother's house with proprietorial fascination, hoping our visits would coincide with its brief season of splendour.

Here, it sheltered a colony of gaunt hydrangeas of a type I'd never seen before, like a chorus line of Giacometti figures. The mulberry was opposite, strafed by wasps, slumping over the centuries until it was almost supine, sprawled lazily on two props. A witch-hazel shared the same bed and across the path there was a corkscrew hazel, *Corylus avellana* 'Contorta', also known by the jaunty name of 'Harry Lauder's Walking Stick'. It formed the backdrop for a spectacular yucca, which had from the day we moved in thrown up spike after spike of ivory-coloured bells.

The yucca was a remnant of what I knew from the photographs in *Country Life* had been a magnificent border, populated with convivial groupings of delphiniums and oriental poppies, their enormous silky heads splashed glamorously with black. Now it looked clogged and unwell. The lawn had spread into it, sending invading tendrils that formed brittle yellow mats, choking the stunted plants at the front. I've never cared much for lawns, but there was something about this sight that crystallised all my feelings of dismay at the disorder. One hot morning I got a hand fork and sat down at the edge, teasing the grass free. The roots formed nets that lifted easily, the one advantage to sandy soil.

As I dug, the tines of the fork struck something metallic. I worried at it with my fingers, and realised it was the old edging, a curved metal line that ran right round the lawn. There was probably nothing more unfashionable at that moment than a neatly edged lawn, and yet to me it was like a talisman, a tangible marker of the balance between disorder and control, abundance and clarity, that every gardener must determine for themselves.

It wasn't all like that of course. There were swims off the beach at Dunwich and Sizewell. On the first night as we went to bed we heard two tawny owls giving their call and response, and every night after they continued their nocturnal conversation.

Sometimes I'd get up and stand at the window overlooking the lawn, a negative dark space beneath a sky seeded with stars. The weather held, unbroken, and every day I slipped out just after dawn, to walk around and drink my tea. That was the best time to be in the garden, at dawn or dusk, when the register of colours was very soft, tinged with pink and lavender and sometimes gold.

I went across the lawn and through the first archway in the beech hedge that led into the pond garden, the space I most loved. It was so private, so remote. Even the sound of traffic was muted. The curved wall was draped in a dense green tapestry of leaves: fig and jasmine, *Akebia quinata* and Virginia creeper, which would soon turn itself a cardinal's scarlet. A banksia rose surged above the rampart, toppling over itself to make a kind of secret passage. There were always birds chittering out of sight, and with them a pleasant feeling of not-aloneness, of being accompanied by small invisible presences.

This was the most formal of the gardens. The quatrefoil pond was at one end and opposite it a stone path ran between two long beds, each with a cypress at the near end, one very tall and the other a little stunted. Both beds were edged with five wobbly mismatched squares of box, and at the far end there was a robinia tree, swagged with candyfloss plumes of blossom. Next to it was a wooden door in the wall, which led into the stables. Above the hedge, you could see the old pigeon loft, capped by a weathervane in the sturdy shape of a Suffolk Punch.

Within this orderly structure, the plants had run riot. At

first glance it was a lovely confusion of colour: palest mauve geraniums, luminous Naples yellow cups of evening primrose and spiked moth-blue heads of echinops. Everlasting sweet peas clambered up the cypresses and there was a trio of standard trained hibiscus trees, pink, blue and white. But apart from one clump of an unusual kniphofia, which I thought might be 'Bees' Lemon', the bulk of the plants were promiscuous self-seeders like Welsh poppy and white campion, or thugs like lemon balm and lamium, which had formed a dense striped mat across nearly every bed in the garden, choking out growth. Something weird seemed to have happened at the stable end too. The soil was much lower there, and there were big gaps between plants. Months later, I discovered this had been the site of another wedding marquee.

The original intention was for this space to have something of the atmosphere and character of courtyard gardens in southern Spain. The unusual quatrefoil shape of the pool, Mark Rumary wrote in *The Englishman's Garden*, was copied 'from those often seen in classical Mediterranean gardens, and doubtless of Moorish origin.' The technical name for this type of design is a paradise garden, and it's far older than Rumary may have realised. Paradise gardens emerged in Persia six centuries before the birth of Christ, and are organised according to strict geometric principles. They must be enclosed and include the element of water, such as a pool, canal or rill, as well as formally arranged trees like pomegranates or cypresses.

These gardens are closely associated with the founder of the Persian empire, Cyrus the Great, about whom Thomas Browne wrote his strange and melancholy tract, *The Garden of Cyrus* in 1658. Over the centuries, they spread across the Islamic world, becoming known as *charbagh* for their fourfold structure. They can be found in Iran, in Egypt and in Spain (Rumary first saw them in the famous Moorish gardens of the Alhambra). In the sixteenth century they were introduced to northern India by the first Mughal emperor, Babur, and though many have since been lost or destroyed, they persist in the form of Mughal miniatures: exquisite spaces with terraces and pavilions, planted with distinct, recognisable flowers like irises and lilies, popu-lated by birds and fish and busy retinues of gardeners.

At first I thought it was the gardens that were being compared to paradise, as in heaven, but to my surprise the concepts ran the other way round. Our word 'paradise', with all its charmed associations, has its roots in Avestan, a language spoken in Persia two thousand years BCE. It derives from the Avestan word *pairidaēza*, which means 'walled garden', *pairi* for 'around' and *daiz* for 'build'. As Thomas Browne explains in *The Garden of Cyrus*, it was these botanically-minded people 'unto whom we owe the very name of Paradise; wherewith we meet not in Scripture before the time of *Solomon*, and conceived originally *Persian*.'

The Greek historian and military leader Xenophon of Athens encountered the word when he was fighting in Persia with Greek mercenaries in 401 BCE. It first appeared in the Greek language in his description of how Cyrus the Great planted pleasure gardens wherever he travelled: παράδεισος, transliterated as *paradeisos*. It was this Greek word that was used in the Old Testament to refer both to the Garden of Eden and to heaven itself, irreparably entangling the celestial with the terrestrial. From there, it migrated on into Latin and then to many other languages, including Old English, where it gathered further meanings. By the thirteenth century it had also come to mean 'a place of surpassing beauty or delight, or of supreme bliss', which is to say that having ascended to the sublime, it returned to earth again.

Discovering this chain of associations blew my mind. It was the garden that came first, heaven trailing in its wake. That had been the acme of perfection, the ideal across centuries and continents: an enclosed garden, a fertile, beautiful, cultivated space. I loved that the material predated the sublime, or rather that the sublime arose from out of the material. It upended the creation myth in a way that was intensely pleasing. I'd once heard it called the English heresy, that paradise can be located in a garden, but actually it was where the rumour of paradise was founded.

Perhaps it also struck a chord with me because my own first encounter with the concept of paradise was so entangled with a real garden. I was educated at a convent in the Buckinghamshire village of Chalfont St Peter. The school occupied a very old house called the Grange. The girls said it had belonged to the Hanging Judge Jeffries, whose bloody assizes still lingered in local memory. It was set inside substantial gardens, encircled by a dense palisade of conifers. There was an orchard behind the tennis courts, and here and there you'd come across statues of Jesus and Francis of Assisi, their concrete palms upturned.

My teacher in kindergarten was Sr. Candida, still the most truly gentle person I have ever met. Devotion was her permanent atmosphere, a kind of travelling cloak. She spoke softly, was never unkind, and as an escape from the classroom took us on nature walks through the woods. It was there I learned to recognise *Vinca major*, periwinkle, with its striped yellow leaves and clear blue petals. Once we were promised a surprise. We clambered down into the Dell, a previously unguessed at space. Set into the bank was a flint grotto and inside it a small painted figure of the Virgin, her head bowed, cradling the infant Christ.

Catholic kitsch, but it got under my skin. We weren't always taught by nuns, but stories from the Bible seeped into everything. I remember puzzling over the Creation, aged five or six. It didn't fit with the more scientific narrative I'd also been given, and so I concluded with what was definitely not complete satisfaction that God must be responsible for the more difficult bits, like eyelashes. I had a nightmare around the same time, in which the word *furnace* was used, one of a species of word dreams that continue to today. I understood it meant burning and woke terrified. Hell below, paradise above, and somewhere off to the side the Garden of Eden, which seemed to exist in many forms, including the convent garden itself, secluded and ruled over by a mysterious hierarchy of spiritual beings, as well as every transformational garden of my childhood reading, from Greene Knowe and Misselthwaite Manor to the midnight garden where Tom sports in the tall trees.

I didn't like the Bible stories I was told, too didactic, too cruel, but I did like the strangeness of the second creation story in Genesis, in which all the plants exist, pre-formed, beneath the earth and God makes man and then plants a garden and orders him to dress it and keep it. Here was a theological commandment I was willing to get on board with. Later I loved the Song of Solomon for the same reason, the hypnotic language of the King James version, weaving around repeating metaphors of sealed fountains and orchards, green apples, green figs, cedar trees, lilies.

Eden was precise; it existed in a way that heaven didn't, even if its gates were locked. I sound like a young heretic, which I was, but my sense of it as a somehow real, tangible place was shared by centuries of religious thinkers. The question of Eden obsessed the medieval imagination, existing as a kind of paradise without, a more imaginatively and physically accessible, maybe even alluring version of the abstract domain of heaven. Where exactly was it? What kind of plants grew there? Could it be located on a map?

I once spent a day sitting by a broiling radiator in the Warburg Institute in London, looking at depictions of Eden drawn from medieval Bibles and romances and books of hours. In one of the strangest, it was a tiny island, surrounded by ocean and irrigated by four rivers, presumably Pison, Gihon, Hiddekel and Euphrates, which cut into the land most dangerously. Men in boats like walnut shells – coracles, I suppose – were approaching by way of each of these swift channels. The island itself was walled, with archways to let the water through, and inside these walls was a single fountain, the sort of thing you might find in any Roman square, which fed the rivers by way of four lion's mouths. There were yellow flowers, which looked like Solomon's seal, and outside the walls two trees, at the foot of two tiny mountains the size of molehills.

Countering this spartan and stylised vision was Cranach's Eden, which like the Mughal miniatures teemed with real, recognisable plants and animals, from pheasants to fat bears. Everything was happening at once, in an orgy of simultaneity. In the background you could see Eve being tugged bodily out of Adam's side, her legs still buried in his viscera. On the other side of a pear tree, the loving couple were tucking into the forbidden fruit, urged on by a devil who was half serpent, half child. God was in many places, watched sidelong by the quail and the grazing deer. There were herons and an angel with a sword, but for me the central enchantment was the inordinate detailing of the green and fruitful world, back when it was new and unpolluted.

What Cranach managed to convey in a single frame was the dual nature of Eden, as both a place of plenitude and pleasure, and as a site of transgression, whose inhabitants are brought into being only to find themselves sent forth and driven out for the crime of disobedience, for which their entire species is cursed (in early nineteenth-century England, the peasant poet John Clare was still ruing the hardships of farming as a consequence of Adam and Eve's temptation). This expulsion too forms part of the powerful allure of Eden, in that it functions as a lost place, where all needs were met and pain was not yet invented, a prelapsarian paradise that stirs longing both for an unpolluted ecology, an intact, immaculate landscape, and the pre-birth heaven of life in utero, where hunger and separation are unknown.

Absurdly, it is only in writing these words down that I realise this aspect of Eden too was mirrored in the convent garden. In those years, my mother was in a closeted relationship with a woman, and when I was nine she was outed among the parents at the convent. It was impossible to stay in our village or school and we ended up moving to a new town hundreds of miles away, where we knew no one. I never saw the garden again, except once, decades later, when I went back with my friend Tom. We were wandering through the Dell and a nun stopped us and looked me in the face a moment and then greeted me by name. The apple of knowledge had been plucked, and maybe the knowledge was sexuality, or maybe it was the knowledge of how cruel the policing of sexuality can be. Either way, the garden gates were barred.

Many people have lost a paradise, and even if they haven't the story of the lost paradise continues to resonate because nearly all of us have lost or relinquished or else forgotten the paradise of a child's perception, when the world is so new and generous in its astonishments, let alone the sweet, fruitful paradise of first love, when the body itself becomes the garden. Perhaps this is why literature is so crammed with secular versions of the Eden story: gardens that open unexpectedly and are then locked, a paradise that is stumbled upon and can never be found again; 'that low door in the wall', as Charles Ryder puts it in *Brideshead Revisited*, 'which others, I knew, had found before me, which opened on an enclosed and enchanted garden.'

Both *Brideshead* and its melancholy French equivalent, *Les Grande Meaulnes*, centre on gardens of dangerous enchantments, a taste of which can warp or blight a life, making everything else seem threadbare by comparison. This type of garden, though, tends to be more aristocratic than divine, a paradise of idleness and luxury, where the springs have turned into baroque fountains, and a masked Adam feasts on plovers' eggs and strawberries, washed down with champagne.

These were the kind of thoughts that moved through my head as I worked, drifts of memories and ideas that were at once tethered and stimulated by the work my hands and eyes were doing. It was another kind of paradise altogether, to be working physically, hours every day, at something so immersive and all-encompassing. Each morning after breakfast I gathered up my tools: red secateurs, hand fork, red plastic bucket, sometimes a wheelbarrow or fork or pruning saw, often a rubble sack. I went to wherever caught my eye, and worked over it, just as an editor rakes over a text, looking for what doesn't belong; what impedes the view or breaks the flow. Often I didn't stop until the light faded, and then at night I read books about gardening, by Christopher Lloyd and Margery Fish and Russell Page.

One morning, my task was taking couch grass out of the little box parterre. I crouched on the lawn, teasing roots out with my fingers, disentangling the silky threads from the parched leaves as swifts screamed overhead. Another day, a fortnight after we moved in, it was the ropes of bindweed that snagged my eye. They had invaded the last of the gardens, which was screened from the rest by a shaggy yew hedge. You entered it two ways, either through a collapsing brick archway, held together by a Gordian knot of ivy, or alternatively between the two fastigiate yews that grew unpruned beside the old coach house, boughs sticking out at rakish angles, like unset bones.

Inside, everything had gone to rack and ruin. The original vision of white-toned, shady clarity and coolness had long since been lost. One of Mark's quartet of corner fruit trees had vanished altogether and another had collapsed. The third, a winter-flowering cherry, was half-dead, forming a scaffold for a rambling rose. A thicket of bamboo nearly blocked the path and the north wall was no longer decked in the linen-coloured lace-caps of *Hydrangea petiolaris*. Instead the grey gault bricks were concealed behind a shifting curtain of bindweed, which sent its tendrils up the medlar and into the abandoned garden next door.

I began to peel it off the wall, grabbing the green cords and hauling them in, releasing great showers of dust and cobwebs. It lifted much more easily than I'd been expecting and soon I'd exposed more algal blown bricks. The roots were a different matter. Just beneath the surface there was a seething network of fat white stolons, each as thick and juicy as cooked spaghetti. They were intertwined with the tougher spiked brown roots of bamboo, which could regenerate from the smallest fragment. I broke a hand fork rootling them out and carried on with a heavy garden fork, sweating and swearing as I levered them up.

There's something intrinsically satisfying about this kind of work, which feels as if it is psychic as well as physical labour, somehow deeper or more primitive than simply tidying a garden. It echoes the kind of tasks set in fairy tales, like the impossible challenges Psyche is given, where she must sort grains of dirt from poppy seed. By noon the bed was clear, and there was a heap of roots and dead branches nearly as tall as me. I shoved it all into a blue Ridgeons rubble sack I'd found in the shed, and finished by cutting back the bamboo, snipping it into what I hoped were vaguely curvaceous, bulbous shapes. The stump of the missing cherry emerged, along with dozens of hellebores. There really was a garden under here. It could be retrieved.

I thought about the convent often in those weeks. It came back as whole cloth, the feeling of security, the contentment of that period. I hadn't ever really felt that sense of absolute, unquestioning belonging again, though each garden I'd made was a snatch at recovery. I knew the nuns had sold the convent not long after my final visit, and then become embroiled in a long-running legal dispute with the council over whether it could be turned into a housing estate. One evening, I idly looked up the Grange on the internet, and so discovered for the first time how closely it had intersected with the established history of paradise.

Alongside many local news items with clickbait headlines like 'Judge rejects claims by parish that nuns conspired', I found a scattering of older stories. In the seventeenth century, the house had belonged to Sir Isaac Penington, Lord Mayor of

London, a Puritan who fought for the Roundhead cause in the Civil War and then served in Oliver Cromwell's government. After the restoration of the monarchy in the spring of 1660, he was imprisoned in the Tower of London on the charge of high treason for his role in the execution of Charles I. A few years earlier, he'd given the Grange to his son, also called Isaac Penington, as a wedding present. The younger Isaac and his wife Mary became enthusiastic Quakers, turning the house into a centre for the exchange of Quaker ideas.

In those turbulent days Quakerism was considered acutely dangerous and subversive, even seditious, for its rejection of any rank or authority aside from that of God. I read on, fascinated by tales of local Quakers persecuted and arrested for refusing to say oaths or doff their hats. Many of these stories came from *The History of the Life of Thomas Ellwood Written in His Own Hand*, the autobiography of a squire's son from a nearby village, who joined the Society of Friends after hearing Quakers preach at the Peningtons' house. His father tried to prevent this politically perilous course by beating him and stopping his allowance, and so he escaped to the Grange, eventually becoming tutor to the Penington children. In 1662 he caught smallpox and after his recovery moved temporarily to London. There he served as an amanuensis to a friend of the Peningtons, the blind poet John Milton, who was living at the time on Jewin Street, where the Barbican is now. Like Isaac senior, Milton had served the Republican cause and was still fearful of reprisal for his support of the execution of Charles I.

It was Thomas Ellwood who found Milton the cottage he

called 'a pretty box' in the neighbouring village of Chalfont St Giles, a few miles from the Grange, when the plague of 1665 made life in London too dangerous. And it was there that Milton showed him the new poem he'd just completed. He'd written all 10,550 lines of *Paradise Lost* without the faculty of sight, hearing the verses like dictation while the rest of his household slept, and waiting in his chair for an attendant to come, so that he could be 'milked', as he put it, of the night's accumulated language.

How strange time is. The same patterns keep recurring, a helical sequence of war and sickness, from out of which emerges the same green-veined dream. Eden floated above the garden of Milton's cottage like a transparent ball; a contained world, rising and falling on the warm air, just as the convent garden bumped sometimes at the corners of my vision. No doubt I would have come to the poem anyway, but this coincidence in place and circumstance brought me more smartly to *Paradise Lost*, which for all my Edenic voyages I had not yet read. I opened it for the first time in a new season of plague, immured in my own garden; frightened, as Milton was, by a sickness that moved invisibly, spreading from city to city – 'the pestilence then growing hot in London' – so that you winced and turned away when a stranger coughed.

All I really knew beforehand was the odd phrase, like 'his dark materials' and 'darkness visible', and bits of theology that had washed up in other books. A novel called *The Vintner's Luck* introduced me to the war in heaven and its rebel angels, while Philip Pullman's trilogy borrowed much of its architecture and imagery from Milton. But none of this prior knowledge prepared me for the labour or the thrill of reading *Paradise Lost*. The story itself wasn't unfamiliar, since it is largely a reworking of the books of Genesis and Revelation. But Milton liberally embellished and expanded his source material, creating an epic around the circumstance of Man's fall.

A civil war has caused Satan and his crew to be expelled from Heaven and imprisoned in the penal colony of Hell. We first encounter them cast into a fiery ocean, bobbing unconscious on the waves like survivors from a terrible shipwreck, still furnished with their weapons and martial banners. The air hurts to breathe, and reeking fires burn but cast no light. Ashore, in darkness and constant pain, they construct a fantastical palace and plot revenge on God, the architect of their downfall (like politicians, the devils never think to blame themselves). It's proposed that Satan seek a rumoured new world, 'the happy seat / Of some new Race call'd Man', and there seduce and ruin God's most recent creation, by brute or subtle means.

I'd never read anything as kinetic as Satan's journey across the universe, a prison break of cosmic proportions. Grieving and embittered, he travels through a chaos without dimension, which seethes and shifts its form, composed of mixed elements perpetually at war, now a burning quicksand, now an icy cold vacuity. The lines throb, so that you feel compelled to mutter them aloud. 'With head, hands, wings, or feet pursues his way / And swims or sinks, or wades, or creeps, or flyes'. At last

Satan glimpses the Earth, in that loveliest of all Milton's images, fast by Heaven like a pendant in a golden chain, as large as the smallest star glimpsed beside the Moon.

I was less interested in the negotiations between God and the Son in Book 3, keener to return to Satan and see with him the new planet, Earth. He finds his way there by disguising himself as a minor angel and begging directions from Uriel, regent of the Sun, who foolishly points out 'Paradise, *Adams* abode'. Unobserved, Satan voyages to the border, a plateau on a steep hill planted with giant pines and cedars, surrounded by a verdurous wall. Somehow higher still are the fruit trees, magically hung with blossom and fruit at once (part of the pre-Fall nature of Eden is that it's in a perpetual state of fecundity, that illusion gardeners strive for still). The air is perfumed and delicious, 'pure now purer', a painful counter to the unbreathable air of Hell, and Satan climbs pensively upward, looking for an entrance.

You could get drunk on the language here, as it swoops between classical and earthy registers. When Satan scorns the gate and leaps the wall: 'So clomb this first grand Thief into Gods Fould', a line I took to quoting when I entered the kitchen in search of snacks. Landing, he finds himself inside a true paradise, heaven on earth remade. The flowers aren't planted in 'Beds and curious Knots', like the fashionably artificial gardens of Milton's own time, but instead grow abundant and unchecked, in open fields and densest shade alike. There are groves and lawns, caves and grottoes, trees, fountains, rills and glades, and everywhere the trembling air smells of the fields, of ripe fruit and freshly tedded hay. Eden is not at all the cultivated, stylised garden of the medieval paintings I'd seen at the Warburg, but instead a wild place, at once delicious and grotesque, perhaps inspired by Milton's youthful travels through the Renaissance gardens of Northern Italy.

The two humans who inhabit this pristine wilderness are naked, eating a supper of nectarines, surrounded by frisking wild animals, united and non-hostile. This is the vision William Blake and his wife Catherine sought to recreate when they played at Adam and Eve in their garden at Lambeth, shucking off their clothes and sporting in the sun. But what struck me most about Milton's pair is that they're not idlers in paradise. They might make love in a bower carpeted with wild crocuses and roofed with climbing roses, but their job is 'sweet Gardning labour', and it occupies the magnitude of their energy and time; a source of anxiety as well as joy.

There's a distinct note of bathos to Adam in particular, worrying like a suburban Sunday gardener about keeping Eden's paths clear of unsightly blossom. Margery Fish's tidy-minded husband Walter often cuts a similarly fussy figure in her gardening books, armed with his broom and secateurs, rooting out any plant that dares to set seed in a path or lawn. As he prepares for sleep, poor Adam is already fretting about the need to rise at dawn, 'to reform / Yon flourie Arbors', which will need cutting back lest they become impassable. Eve too says, though with less concern: 'What we by day / Lop overgrown, or prune, or prop, or bind, / One night or two with wanton growth derides / Tending to wilde.'

Eden always exceeds their efforts, is perpetually in growth, lavish, excessive, profligate, wanton, intransigent, abundant. Their task is to tend it, not by hacking and levelling, but by the gentle arts of pruning and training. Since forging iron is a post-fall invention, their tools are rudimentary, 'guiltless of fire', but basically their days were spent a lot like mine. In fact, who was I to mock? *Immensely satisfying*, I'd written in the first few pages of my diary, with regard to sweeping the paths clear. I did it nearly every day, as an antidote to the disorder elsewhere.

I hadn't expected the poem to be so relevant to such specific gardening concerns. In another acutely recognisable scene, Eve lays herself open to danger because of her insistence on working unaccompanied in a favourite spot, a concealed grove of roses she's been sculpting and shaping. Too real! She pleads with Adam to let her go there on her own, arguing they'll get more done in separate spheres, without the distractions of 'casual discourse', aka chatting. Plausible, but I suspected what she really wanted was to be alone, the only possible way to dissolve back into relation with the vegetal world (Milton's friend and colleague Andrew Marvell says it even more plainly in 'The Garden': 'Two Paradises 'twere in one / To live in paradise alone'). Adam, on the other hand, prefers to work in close, if not claustrophobic proximity. He claims this is to guard against the danger of temptation, but it's clear he doesn't share her unsettling intimacy with the garden.

During the pandemic Ian and I often argued about territory and space, and the worst of our arguments took place in the garden. It was a place where I did not want to speak, where being made to answer questions or discuss plans felt as intrusive as someone yanking on my hair. I didn't want to be alone, as in on my own, so much as immersed in the non-human world, absorbed by way of silence. It was like entering a pool of water and swimming quietly there, though it's true that often I was muttering to the plants, telling them what I was about to do, or singing some idiotic song that had welled up from childhood, including, on one occasion, the *Flintstones* theme. Either way, I barely thought in conscious terms. My eyes saw something, and then my hands were there, working away. My senses lay open and my thoughts drifted freely, left to their own devices, sluicing more or less inaudibly, sometimes turning up unexpected correspondences or voyaging far back in time. Being pulled out of this state was painful, and it made me incandescent to be disturbed.

I had a suspicion Milton wasn't totally unsympathetic to this aspect of Eve's nature, even if her disobedience did have fatal consequences. Her relationship with the garden occasions some of his most tender writing. In the final scene before the Fall, it's as if she's merged or fused with the plants around her, so that she appears weirdly veiled in fragrance, 'half spi'd, so thick the Roses bushing round / About her glowd'. Carefully, she ties them up with myrtle bands, supporting their drooping heads. The garden here is envisaged as a place of mutual support and interconnection, so delicately balanced that a single missing prop spells disaster. What has escaped Eve is that she is also part of this fragile ecosystem, an unsupported flower and a storm ahead.

That September the garden was full of surprises. I cut back thickets of honeysuckle and discovered astrantia, known as melancholy gentleman for its stiff Elizabethan ruffs and odd, pinkish-green livery; cut back the dead leaves of a fern and found the yellowing leaves and flushed pink stems of two peonies. The rose bed was filled with a confounding excess of plants, most visibly a pompom dahlia with red and white stripes that I thought might be 'York and Lancaster', which seemed to thrive in the most hostile conditions. Once I'd levered my way in I saw there were far more roses than I'd realised, including a deep crimson Portland rose, which dates from 1750 and looked half-starved. At the opposite end of the border, three sedums were pressed up against a clump of crinum lilies, a red-hot poker and another peony, a box spiral, an oriental poppy and a standard trained wisteria, all in a space about the size of a double bed. The dahlias I evicted into a cardboard box to overwinter in the shed, and the rest I mulched. Watch and wait.

One day I found what I thought was a crocus. The next morning there were dozens of them, elegant little ghosts the shape and size of wine glasses, their stems pallid and their bowls a milky mauve. They were colchicums, also known as naked boys, which flower in the autumn, vanish away, and produce transient leaves in the spring. That explained the masses of yellow bulbs I'd found, just beneath the surface of each bed. Some of the colchicums were pure white and a very few had such an abundance of petals that they looked like water lilies. Meanwhile, the bed under the mulberry that had seemed so unpromising filled with a wash of pink and then white cyclamen, hundreds upon hundreds of them. Evidently Eden had survived in pockets after all.

I'd been reading *Paradise Lost* very slowly. The poem was so seamed with oddities and digressions, on the nature of angelic bodies and the movements of the planets, so richly stitched and embossed with detail that it took me a while to identify the moods that seemed to move through it. As I read I often experienced currents of recognition, not so much at the plot as at the kinds of deep emotion that kept seeping up. The interminable classical allusions were like a thicket you had to climb over or press yourself through, and on the other side were zones of pure feeling that had somehow overflowed the story, or fed it like wellsprings.

There was Satan weeping in Hell and Eve weeping in Eden. There was Adam raging at his wife, and the fallen angels seething with bitterness and jealousy, until they were all turned into hissing snakes. Everywhere plans had gone awry: revolutions defeated, relationships severed, beloved homes destroyed. A lava of grief, fury, despair bubbled just beneath the surface. The topography was utterly foreign and yet emotionally it didn't feel like an unfamiliar landscape. I'd always assumed the poem was primarily about religion, but reading it that summer, in a global pandemic, after years of political instability, it seemed to me that it was just as seriously concerned with failure. In fact, wasn't that why it started in the middle, in medias res, rather than with the rebel angels plotting their revolution? From the moment Satan awoke on the burning sea, it was a sustained encounter with failure and especially with what happens after failure has occurred.

Looked at through this lens, the poem assumed a different shape. In effect, it was composed of two acts, which mirrored one another. First Satan's disobedience caused his banishment from Heaven, and then Adam and Eve's disobedience caused their banishment from Eden. All three suffered enormous, intolerable losses, plunging them into a wasteland of negative feelings, a terrible, damaging encounter with shame and rage. Satan is corrupted by his failures, and seeks a multiplication of suffering as a panacea, while Adam and Eve are humbled by what they've done, choosing to relinquish what they had rather than create more damage and pain.

If you were to make a taxonomy of the types of loss contained within the poem, much of it would be concerned with exile: the grief of losing a homeland or a once great estate, of realising that what you thought was yours is not, and that something loved is ruined utterly, gone for good. Eve in particular gives voice to unbearable desolation when she realises her punishment is not just death; that she is also required to leave her home and will no longer be permitted to tend the roses she has named and watered. Her luminous collaboration with the garden is over. From now on, part of her sentence is to be estranged and alienated from nature, tilling weeds in infertile ground, a desperate future we were heading towards ourselves.

A modern reader can hardly help but feel outrage at the engineer of all this unhappiness. After I finished *Paradise Lost*, I followed it with *Milton's God*, William Empson's impassioned attack on the authoritarian and bloodthirsty Christian God, published in 1961. Empson points out that for all its pure air and abundance of fragrance, Eden is little better than a prison camp, a flowering panopticon watched over by angelic armies, with their surveillance patrols and eviction units. God is the great manipulator, hoarding information, keeping his subordinates in the dark, insisting on total obedience in authoritarian trials of blind faith. Eden is a place of paranoia, where no one knows exactly what's permitted, or what kind of knowledge can be acquired or shared. When Adam and Raphael engage in their long and ranging conversation, which covers among many topics the war in heaven, the circumstances of the creation and the nature of angelic sex, both speak haltingly, hesitantly, uncertain of what might be forbidden to discuss.

No poem could be less reducible to autobiography, but the knowledge that certain kinds of speech could be dangerous, even lethal, was shared by Milton himself, as was the sense of being overtaken and implicated in a catastrophe. He wrote *Paradise Lost* at a time of great peril, as the extent of the failure in which he was enmeshed was rapidly becoming clear. He'd been on the Republican side – on what he'd believed unequivocally to be God's side – in the English Civil War, and had watched in horror as the solid fruits of victory vanished into thin air. In just over a decade the world had turned upside-down twice over: a king beheaded and a new king crowned, with regalia that had to be made from scratch, since Cromwell had ordered it melted for coins.

Though he was never a soldier, Milton was among the most

vocal and eloquent supporters of the English revolution, a poet who plunged himself into political life. He'd written tracts calling for and defending the execution of Charles I, argued in favour of Church reform, divorce, Republican government and religious tolerance (though not, of course, for Catholics). The Interregnum was the great era of tract writing and tract reading, when every day new causes and ideas were hammered out and discussed all through the City thanks to the newly enlarged freedoms of the press, another of Milton's many causes. His writings were widely read in Europe too, forming the central scholarly defence of the Parliamentarian cause. After the war was won in 1649, he was asked to join Oliver Cromwell's government as Secretary for Foreign Tongues. Three years later, when he was forty-four, his failing vision became total, so that he sat blind and attentive at the ringside, a witness to the creeping corruption of what had once been radical ideals.

It isn't clear exactly when he started *Paradise Lost*. According to John Aubrey's *Brief Lives*, Milton began it two years before the king's return, working on it only between the autumn and spring equinox, and taking in this way four or five years to finish. His habit was to dictate and be read to from early in the morning until dinner, after which he walked led by the hand for three or four hours in his garden (he always had a garden, Aubrey says), or else sat in his doorway, wrapped in his grey coat. This time frame means that he started the poem as the Republican government began to founder in the wake of Cromwell's death in 1658 (the same year, it strikes me now, that Thomas Browne published *The Garden of Cyrus*).

He was probably halfway through it when the ground crumbled beneath his feet. In 1660 Charles II returned from exile. On 29 May he rode in triumph into London, preceded by troop after troop of Royalist soldiers, in scarlet and gold livery, or sea-green laced with silver, bearing their standards and followed by what the diarist John Evelyn claims was twenty thousand men on horses. Milton had gone into hiding at a friend's house on Bartholomew Close, certain that he would soon be hanged for his support of the king's execution. He was so near the route to Whitehall that he could smell the bonfires and hear the cannons and fireworks and tipsy shouts of 'God Save the King', as the procession wound for seven hours through the city.

Back in April Charles had promised a general pardon for acts committed during the Civil War, with the exception of those directly involved in the execution of the king. At around the same time, though, Milton recklessly republished an enlarged and revised version of his pamphlet fulminating against the monarchy. He was a sitting target for retribution, his blindness paradoxically making him dangerously visible.

For three frightening months Parliament debated who would be exempted from the Indemnity and Oblivion Act and punished as regicides. Almost every day names were added and struck from the list. An order for Milton's arrest was issued in June and in July copies of his books were burned by the public executioner at the Old Bailey. But when the Act passed into law at the end of August, his name was not on the list of regicides and their supporters exempted from the pardon. Even now, no one is quite sure how he escaped. It's possible that his friend Andrew Marvell interceded on his behalf.

For reasons that remain unclear, he was imprisoned for several weeks in the Tower of London, securing his release with a hefty fine. He spent the autumn lying low in his own house on Jewin Street, fearful of assassination by newly emboldened Royalists, who thronged the city in drunken mobs. Frankly, he was lucky to be alive. That October ten of the regicides were hanged, drawn and quartered as traitors. The diarist Samuel Pepys witnessed one of the executions and heard the watching crowd give a great cheer of joy when the bleeding head and heart were held up. A few days later John Evelyn recorded the death of four more 'murderous Traytors', writing, 'I saw not their execution, but met their quarters, mangled & cut & reaking as they were brought from the Gallows in baskets on the hurdle.' Their severed and parboiled limbs were nailed up on Aldersgate, around the corner from Milton's house, as a warning to future rebels.

He was, as the novelist Rose Macaulay puts it in her biography, 'moving bitterly among the shattered fragments of a Republican poet's dream.' At the start of Book 7, the narrator of *Paradise Lost* appears to address these parlous conditions directly, describing himself:

. . . fall'n on evil dayes
On evil dayes though fall'n, and evil tongues;
In darkness, and with dangers compast round,
And solitude;

Some people say that communism has never been attempted and I wonder if Milton would say the same thing about the ideals for which he'd fought. The Good Old Cause was not just about installing one type of government over another, or even ending an illicit tyranny. It was an attempt to found God's republic in England, to build a new paradise on Earth, and it regarded the setting up of kings over men as an intolerable blasphemy. To have Charles II in his robes of insulting scarlet walk up the nave of Westminster Abbey and be crowned by bishops was to feel as if God had turned and spat in the rebels' faces. That humiliating. That personal. What if Milton had been wrong? What if his vision was corrupt? How should he proceed? Should he stay stalwart in the testing fire, or should he atone?

The poem imagines what it's like to inhabit this state of failure and disillusionment, which is shared by all three major characters, but it also considers how to proceed from it. It's propelled by an almost intolerable need to understand what it means to have failed and what one ought to do once failure has occurred, both by imagining a process of future reparations and by re-envisaging the nature of an intact, untarnished world. Maybe this was why it felt so compelling to read in our own season of turmoil. My abiding sense of the past five years was of waking in the dark to check election results on my phone. First there was Brexit, and with it the global rise of the far right. Then Trump was elected president, and every single day more cherished assumptions were crushed into the ground. Concepts like democracy, virtue, truth, liberalism became the punchline of Twitter jokes, then the subject of attacks in newspapers and by MPs. I stopped being able to write or even speak sometimes. Everything I wanted to say sounded exactly like the sort of thing a person like me would say, a stupid liberal, still entranced by dead ideas. It was like the hollowing out of a loved body, head and heart held up bleeding to a laughing crowd. This was part of why making the garden felt so compelling to me: to repair something, to make something beautiful without having to engage in the realm of words at all, a compulsion I described in my diary of the time as *desperate*.

On it went: parliament prorogued, politicians assassinated by fanatics, and now a new plague, with mass graves and the wealthy fleeing to the countryside. It wasn't hard to catch the echoes rebounding from the seventeenth century. So much for the end of history. And yet Milton didn't give way to despair, or relinquish his ideals. He sat in that turbulent and fearful darkness and imagined the mother of all utopias, Eden, right down to how the air might have tasted. 'A Wilderness of sweets . . . Wilde above Rule or Art.' What a strange, improbable thing to do. In my own time there was a great inclination to imagine dystopias, as if the shift to cruelty and authoritarianism had inculcated a desire to pre-empt, to know the next bad step before it occurred. Milton didn't think like that. It was as if he needed to establish Eden securely, inch by inch: a memory palace of a better way of being, a realm of interdependence and mutual support that had been lost but could perhaps be founded again.

Like all utopias it was comprised of elements that were unavailable or scarce in daily life, as well as being assailed by known threats. He dreamed of it as hospitable and abundant, pristine and unpolluted. Air that could make an angel reel as opposed to London's foul miasma, so contaminated by the burning of sea coal that it coalesced into what John Evelyn described as 'a hellish and dismall cloud', through which you could hardly make out the face of a neighbour jostled against you in a crowd. The real city Milton inhabited was closer to Pandæmonium, the capital of Hell constructed by the devils, where it rained sulphur and noxious fires burned in the dark.

Eden is by contrast a place of natural harmony, where resources are infinite and diversity is a source of delight, not threat. The leopard and the stag, the 'Parsimonious Emmet': all rise kicking from the soil, born bodily from the earth. This innate wildness had a religious significance, too, symbolising an emphatically Protestant vision of right relationship with God: direct and natural, without the intercession of a corrupt and elaborate Church. Adam's spontaneous song of prayer, naked and outdoors, is a more bucolic manifestation of the same system of belief that prompted Cromwell to stable his horses in Ely cathedral, to shit in the transept and the nave.

But what mattered more to me was that despite its wild aspects, Milton's paradise was a garden. He used the model of horticultural cultivation as a way of considering good government, right relationship between different kinds of beings. Eden runs on very different principles to the autocratic rule of heaven. If God lays down laws and punishes transgressions, Adam and Eve practise a style of custodianship that is under-stated and benign. When they prop a rose or twist a vine into an elm, they are active collaborators in an interconnected ecology, subtly stage-managing processes that are already underway in order to maximise abundance and pleasure.

That this blissful state of affairs comes to an end doesn't drain it of its power. Despite its title, *Paradise Lost* is not exactly nostalgic. The garden serves as a kind of lodestar, an experience of nurture and richness that cannot be dismantled and might in future be recreated. Adam and Eve mourn their losses, grieve what won't continue, but when the eviction comes, when the cherubim gather like mist rising from a river, when they are taken by the hand and led to Eden's gate, they look back, drop a tear, and then turn resolutely round. The final line swells with possibility. 'The World was all before them.' Whatever they have suffered, whatever damage has been done, the future lies open ahead.

III

A LANDSCAPE WITHOUT PEOPLE

The first thing I planted was wallflower seed, a variety called 'Blood Red', followed by plugs of 'Fire King'. Miltonic names for a plant that makes a paradise of the driest wall and refuses to respect boundaries, casting its seeds wherever it can gain a foothold. It was plain to me now that Eden was my intention, as it is with so many gardeners, though I wanted to look more deeply, more exactingly at its founding principles, too.

At the end of September the long summer broke, giving way to rain. My stepmother was buried in a downpour. It was a gruesome Covid-era funeral, culminating in the discovery that the sexton had forgotten to dig the grave. We huddled in a pub and came back to a raw brown hole in the ground. The vicar had long gone. I read Auden from my phone in blinding rain. Everyone's hair was plastered to their scalp. We threw handfuls of roses from my father's garden onto the coffin. Two weeks later her will was read, and he rang while I was pruning roses to tell me that she'd only left him a third of their house. It had been bought in her name, and she'd assiduously kept it that way. She was intensely secretive about money, and it turned out that he'd been paying what he thought were contributions to the mortgage into her bank account for more than a decade after it had been paid off.

An era of lawyers began, of bleak *Bleak House* jokes that were never funny and became less so as the case inched nowhere. Since early childhood, my life had been marked by housing instability, but though it was my father who'd set that process in motion by leaving, his own home had remained remarkably stable. He'd lived in his current house for more than thirty years, during which time I'd moved house sixteen times, and had at least as many sublets. We'd moved as children because of divorce, because my mother was outed, because we were escaping her alcoholic partner. In adulthood I'd replicated this pattern, in part because of my own restlessness and in part because of how insecure housing is for a person who rents. All that time, my father had stayed in place. Now he was like a person without a compass, baffled and stricken. He didn't want to go, didn't know if he would be able to stay, couldn't begin to process the feelings that kept leaking out. He'd built the garden from scratch. It was his world.

Old injustices began to surface, packaged in lawyerly language, the emails detonating buried rooms of emotion. I seemed to spend hours on the phone. Get a grip, but I couldn't. Out every day after breakfast, rehearsing in my mind possible responses or refutations. It didn't escape me, what I was doing, the way the garden had become a counter to chaos on a personal as well as political level; where, as I wrote in my diary on the evening of the will's discovery, care is rewarded, a place of emotional investment and security. I felt okay only when I was outside, though as soon as I came indoors the garden too became overwhelming, contaminated in some mysterious way.

A Virginia creeper had appeared like a scarlet sash over the Irish yews, where it had been climbing imperceptibly upward in its summer coat of green. Red for emergency, it drew attention to the trees' sorry state. All fastigiate yews require regular pruning. Without it, they'd grown like broken fans, smacked about by storms, ribs sticking out akimbo. Then there were the cherries in the abandoned garden behind the yew hedge, one definitely dead, the other rotten through.

The tree by the house was even more worrying. I'd finally identified it from one of Mark's essays on the garden: a Lavalle hawthorn, which should by now have been a mass of red berries. Instead, whole sections were bare or hung with desiccated brown leaves. A visitor speculated about honey fungus, pointing out knuckles of dieback in the magnolia too. I fretted through the night and the next morning went out with a screwdriver, to scrape back the hawthorn's bark and see if the tell-tale white fungal growth was underneath. I needn't have bothered. The fruiting bodies were all around the trunk, dozens of spooky golden-brown mushrooms, reeking of honey.

Already in the countryside about you could see the consequences of ash dieback, the thinning canopies and blighted leaves that within a year gave way to skeletal trees. Honey fungus is even more frightening. It feeds off living rather than dead matter, and can decimate whole gardens, wiping out mature trees, shrubs and roses. It's the largest living organism on earth, and can extend for miles under the ground, travelling at the rate of a metre a year. I imagined the black bootlaces of its rhizomes moving implacably beneath the lawn, bringing down the magnolia before encroaching upon the mulberry, which had been growing there when Milton was setting down the first lines of *Paradise Lost*, three hundred and fifty years ago.

The internet did not assuage my fears. The advice from the RHS website was to excavate and destroy all infected material, and then to bury a plastic membrane deep in the soil to stop the rhizomes spreading. What, dig out and burn two trees? I turned to gardeners' forums. People suggested using Jeyes fluid as a substitute for the banned poison Armillatox, or cutting down the affected tree and leaving the area unplanted for three years. There was a barely concealed invasion narrative in all this, of bad actors and weaponry so dangerous it needed health warnings. It was as if the garden had become a battleground, swarming with enemies that needed to be expunged, and that furthermore could be without causing further damage. I was frightened by the fungus too, but I didn't want to pour tar acids and chlorocresol into the soil, wiping out micro-organisms and insects, while removing a tree in the hope of depriving the fungus of a host made no sense in a small garden surrounded by dozens of others. Wasn't it a network, after all, a self-organising whole? It's not how I want to garden, I wrote that night, or be.

But it wasn't just the trees that were bothering me. Ever since I'd spotted the wooden roof struts in the greenhouse were rotten I'd entered in trepidation. The wall at the back was sooty with mould and the jasmine growing from a planting hole by the hand pump was infested with mealy bugs. The pump, of course, was broken. There was such an impatience in me to mend these things, to press my shoulder against the garden and push it into what I could see so clearly, its future shape. It's amazing I've ever managed to do anything as necessarily slow as writing a book. I want an idea to exist fully-formed the second I've sketched it out. At the same time I was grateful to be faced with so much that needed mending. I needed to be sorting, cleaning, fixing, making something orderly out of disarray, as if by doing these tasks I could assuage an almost unmanageable anxiety.

For these particular jobs, though, I needed help. The first person to appear was the tree surgeon a neighbour had recommended. He wandered in as we were drinking our tea, lanky and shambling. A gardener lived here, he said, and looked pleased when I knew who he meant. We inspected the trees together. There was a great gouge out of one, like a slice of cake. Presumably a branch had torn away and in response someone had roughly hacked out nearly half the trunk. Lee ignored what seemed to my eye like mortal damage and looked up instead. It was an ornamental pear, he said, and would live for decades more. This was welcome news. The pair of yews he suggested lowering by twelve feet, lest the weight of the damaged branches tear their trunks too. The two cherries, however, had reached the end of their life and needed to come out. They'd cast such dense shade that all the plants below were struggling, listing forward at the same acute angle, like people straining to see. A lilac in particular was so deprived it had coral spot, which might improve given a dose of light and air, like a nineteenth-century patient with tuberculosis.

His phrase stayed with me after he left. *A gardener lived here.* A few days later, the doorbell rang again – an actual bell, which jangled when tugged by a wire. This man was thickset with very blue eyes, and the first thing he said was, I'm on hallowed ground here. Again, a burst of pleasure when I knew what he was talking about. He ran a landscaping business and in his youth had often worked with Mark. We went on another slow circuit round the garden, and he talked me through it as he remembered it, casting disparaging looks at any new and interloping plants. Mark loved pinks, he said, and then, he was gay you know, with a sidelong glance at me to see how I took the news. He cocked a hip and added quickly, but not like that, he was very discreet.

For a moment it was as if I'd fallen back to my own childhood, or even earlier. I wasn't accustomed to being in circles where gayness was an aberration that needed to be explained. Pink. An indiscreet pink flower. But now he was giving me names of people who had known Mark. I scribbled them down, a litany of nurserymen and garden designers, as well as a man in Aldeburgh who he thought had been Mark's executor. He poked at the greenhouse timbers and told me cheerfully it could be patched up.

That afternoon I set about writing cards to all those people. I was eager then for any scrap of information about Mark, who had died in 2010. He fascinated me, in part because of the warmth with which he was spoken about by everyone I met and in part because of the strange intimacy that went along with repairing his design. To mend the garden meant working first or simultaneously to understand the pattern he'd intended, and then to recover it, stitch by stitch, as an embroiderer might. It was like listening very hard to words in a language I only spoke imperfectly. Some of it was conscious, a careful working out from photographs or articles what plants a border had contained in a particular season or year, but more often it was instinctive, reconstructing rhythms and waves of colour, extracting what jarred and gauging what needed to be reinserted to nudge the structure back to life. I often had a sense that I wasn't alone in the garden, as I never did when in the house. Mostly when I had this feeling I would look up and find I was being watched by the glossy black eye of a blackbird or robin. All the same, I was accompanied, pressed up very close to someone else's sensibility, their taste and style.

Small fragments accumulated. An obituary by Tony Venison, the long-term gardens editor of *Country Life*, described a 'quiet, unassuming man, a twinkle of amusement in his eyes', who had won three RHS Lawrence medals for his show gardens with Notcutts, 'an unrivalled achievement' at the time. Mark's Twelfth Night parties, supper tables laid in every room, the gas lamps burning, hundreds of candles, a woman in a dress made riskily of feathers. Trips to Glyndebourne and Bayreuth, the music room full of harpsichords and giraffe pianos, and on the top floor a library of sheet music presided over by a dolls' house. A tortoise called Alphonse, who wintered under the wisteria, and a succession of English setters called Bruno, the last of whom was buried beneath a stone plaque I'd found behind the hornbeams. I gleaned these things from many different conversations, in people's kitchens and on the street in front of the house. Everyone wanted to tell me about Mark, to say, and to say again, what a lovely man he was.

By mid-October the rain had stopped and there was a succession of buttery yellow days. I wandered about, pulling out wood avens and clearing away the limp yellow rags of acacia leaves, unearthing a shining green tapestry of plants, happy and healthy after the long inundation. The witch hazel glowed. There were worm casts on the lawn. I planted the first tulips in pots: 'Marietta', 'Helmar', 'Paul Scherer', 'Estella Rijnveld', with its great crimped petals of scarlet and white. The roadside stalls were piled high with quinces and squashes and in the cottage gardens waterlogged purple asters flopped against the underwater stars of nerines. I was suffused with the feeling of autumn's new book. Clamps of sugar beet began to appear at the field margins and there was an almost hallucinatory clarity to the air.

On one of those bright afternoons the house phone rang. Let me tell you something spooky, a man's voice said. Your card was dated 16 October. I went into the file and found Mark's death certificate and it was exactly ten years to the day since he died. It was the executor I'd been advised to contact for more information. He invited me to lunch, promising to tell me what he described as all the gossip, and adding, Mark was gay, you see, when it wasn't good to be gay.

We went a few days later, for an unexpected feast of blini and smoked salmon. Mark's executor was a large bonhomous man who moved very lightly, jokes and anecdotes tumbling in his wake. He and his wife had become friends with Mark when he designed their garden, which opened onto the marsh, filling the house with watery skeins of light. Among the papers piled on the table was a hand-drawn planting plan, full of things I recognised from home, including the gaunt, fuzzy-leafed *Hydrangea aspera*, the striped crimson and white rose 'Ferdinand Pichard' and that mauvish-blue geranium, 'Rozanne'.

Two more friends of Mark arrived a moment before we did, clutching ferns they'd lifted that morning. They were both painter-gardeners, or gardener-painters, and had studied at Benton End, Cedric Morris's famous art school-cum-garden on the Essex border, where he bred irises in lovely muted ashy colours and ushered a generation of young artists into bohemia. Mark, they told me, was more of a gentleman than a bohemian: the son of a Sussex butcher who studied architecture before finding his way to gardens. He moved in different circles to the Benton End crew. Once the doorbell had rung while Mark was cleaning out the pond, thigh-deep in murky water, dressed only in his underpants. When his partner Derek had opened the door he'd found Princess Margaret and her lady-in-waiting, come to view the garden. Sorry, ma'am, Mark's just putting on his trousers.

He began his working life as the assistant to the American landscape architect Lanning Roper, where it was smart set all the way, and served a spell at Sissinghurst too. During his thirty-odd years as Landscape Director at Notcutts he designed and restored dozens of gardens, mostly in East Anglia, though he also worked for the Jordanian royal family. His contribution to horticulture was so great that in 1995 he was made an RHS Associate of Honour, an award for distinguished service, of which there are only a hundred recipients at any given time. But very low-key, they said, no side to him at all.

One of the gardens they mentioned was Ditchingham Hall, a Queen Anne manor house on the Norfolk border that had been the childhood home of the writer Diana Athill. When I got home I looked it up. It was privately owned and rarely open to the public. The few images I found were pure Mark, lush and romantic: a wisteria tunnel, rare trees, a raised pond tucked into the corner of a walled garden. The name of the house was nagging at me, and eventually I tracked it down to *The Rings of Saturn*. It's the final port of call on Sebald's haunted Suffolk journey, the last station on a long, troubled walk along the coast, in which the narrator, an avatar of Sebald himself, is tugged continually backwards through time, each ruined house or radar station a portal to the past.

I found my copy in the spare room and reread the Ditchingham chapter, sitting there on the bed. His walk that day must have taken him right past our front door, one August morning in 1992. I hoped Mark had been outside, clipping the French fancies. Sebald took the Roman road to Heveningham, detouring to visit a barn where a man had spent thirty years building a model of the Temple of Jerusalem, continually destroying and amending his work as his researches yielded new information, the kind of hopeless, possessed labour that Sebald finds so emblematic and compelling.

I remembered the temple but I'd forgotten that the chapter ended with an apocalyptic vision of dying trees, not just the catastrophic devastation of the Great Storm but the diseases that preceded it in the 1970s: first Dutch Elm disease and then a kind of dieback I thought was the exclusive province of my own century, which caused the crowns of ash to become sparse, and the foliage of oaks to thin and display what Sebald calls 'strange mutations'. The beeches were afflicted too, followed by the poplars, and at last on the night of 15 October 1987 the hurricane came and in the morning fifteen million trees were lying on the ground, as if, Sebald writes, they had fallen into a swoon.

The Rings of Saturn is a book composed of images of destruction on a near-inconceivable scale: the deaths of millions of herrings, say, or eight thousand pheasants on a single day's shoot. The central injury for which all these other injuries are metonymies, parts standing for a whole, is the Holocaust, the near-total destruction of the Jews of Europe. In Sebald's vision the world is and always has been a charnel ground, in which success is at once contingent upon plunder and predation, and doomed to ruin, since the universe will not support perpetual harvest but must pass through cycles of war and famine too. It's a vision of pointless, obsessive labour, including the labour of beautification (that which causes the weaver to destroy his eyes), and of persistent destruction, and so to find within it the fields of mysteriously ailing trees, their leaves turning brown, their capillaries seizing up so that they expire of thirst, was not exactly unexpected, though it startled me to find so close a replication of my present anxieties, of a future filled with dead and dying trees.

At the time of Sebald's visit to Ditchingham, the damage done by the storm was still visible, and it was this damage that Mark was brought in to repair. The great cedar avenue had been partially destroyed, and he was hired to restore trees to the newly naked park, the latest sally in a long campaign to shape and beautify the landscape. As Sebald is at pains to point out, the sublime parkland of the eighteenth-century house might look natural, with its broad expanses of closely cropped grass, its serpentine lakes and pleasant groupings of oaks, but it is a masquerade, a fantasy of what a landscape should be, its calm imperturbability an expensively procured illusion. Contemplating Ditchingham Hall's lavender bricks, Sebald writes:

Estates of this kind, which enabled the ruling elite to imagine themselves surrounded by boundless lands where nothing offended the eye, did not become fashionable until the second half of the eighteenth century. Planning and executing the work necessary for an emparkment could take two or three decades. In order to complete the project it was usually necessary to buy parcels of lands and add them to the existing estate, and roads, tracks, individual farmsteads, sometimes even entire villages had to be moved, as the object was to enjoy an uninterrupted view from the house over a natural expanse innocent of any human presence.

This was the aspect of gardens that had always troubled me: the hidden cost, the submerged relationship with power and exclusion. The work that Sebald describes was known as *improvement*, and judging from Athill's account, the view was terrific, assuming you were fortunate enough to live in the big house and not the discarded village.

In her final memoir, *Alive, Alive Oh!*, Athill recalls the gardens at Ditchingham as they had been in her childhood, in the years between the wars, when the full-dress production of the country house was still tenable, before the Second World War brought an end to the lives in service on which it depended: the teenaged girls who slept like sardines in an attic, boiling and scrubbing other people's sheets, serving up enormous breakfasts of grilled kidneys and home-cured ham and boiled eggs that in all probability no one would even eat. The house belonged to her grandparents, and existed as a kind of contained universe, populated by a revolving cast of family members and a fixed one of servants, among them the cook, the coachman, and the head gardener; a double, apparently, for Beatrix Potter's Mr McGregor, and just as inclined to violence when his vegetable garden was breached.

From the vantage point of her nineties, in a small bedsitter in a retirement home in Highgate, Athill summons all the components of a recherché kingdom: the melon house, the nectarine house, the vinery, the shrubbery. Her itemisation goes on for pages, a dreamlike evocation of a domain as abundant and circumscribed as Milton's Eden. Apple orchard, aviary, even a parma violet frame, all leading up to the central mystery: the walled kitchen garden that Mark later made into such a romantic space. In those days it was bifurcated by a canalised stream and laid out in accordance with a design from William Cobbett's *The English Gardener*, published in 1829. 'It was – it really was – a wonderfully thought-out and maintained fabrication of great beauty,' she writes, adding that its beauty derived not just from its aesthetic qualities but from the garden's robust functionality, 'when its purpose, the sustaining of a household, was real.'

Unlike Sebald, she saw the creation of these landscapes as an act of generosity, not domination.

> The Cedar Walk had been planned and planted by someone who was never going to see it – not him, not his children, not even his children's children, though they would have had a clearer view of what it was going to be. What amazingly generous confidence in the future those eighteenth-century landscape designers had . . . We children . . . were inhabiting a two-hundred-year-old dream: a place planned to give support not only to its inhabitants' bodies, but also their minds – perhaps even their souls.

An idyll, to be sure, assuming you're entitled to be an inhabitant, to walk among the cedars and pick the muscat grapes, your soul expanding in the afternoon light.

For a long time it was wrongly believed that Ditchingham was the work of Capability Brown, the eighteenth-century landscape architect most closely associated with creating the parkland we now experience as so essentially English, though it represents only a single act in the long parade of garden fashions. Brown got his nickname because he saw in every landscape its lucrative capabilities for becoming something it was not. He was among the most famous garden designers to have lived: a Northumberland boy made good, father a land agent, mother a chambermaid, who brought about a wholesale transformation of English taste as well as acres, establishing how we still think a paradise should look.

Brown swept away the French-inflected style of the seventeenth century, replacing artifice with a new naturalism. Out went the hard lines and formal adornments that sound so pleasing now: the parterres, canals, basins, balustrades and bowling greens that had once delighted the gentry. Gone were the cascades spilling and bubbling down flights of steps, the holly and yew trees clipped into fantastic shapes, so that partygoers might wander amongst a crowd of leafy giants, centaurs, milkmaids and sailing ships, or find themselves face to face with Adam and Eve beguiled by the snake, all three of them skinned in dense shining green.

It's ironic that Brown gave Adam and Eve the boot, considering this shift in style drew so deeply on Milton's descriptions of Eden in *Paradise Lost*. The writer and garden-maker Horace Walpole credited Milton as its inventor, never mind that he was already long dead, writing in his influential treatise of 1770, *On Modern Gardening*: 'One man, one great man . . . judged that the mistaken and fantastic ornaments he had seen in gardens were unworthy of the almighty hand that planted the delights of Paradise.' Milton's decision to make Eden a wilderness, a 'happy rural seat of various view', anticipated Brown's wild-seeming lakes and woodlands by a hundred years, articulating a naturalism that would become the height of Georgian fashion.

But these new Edens mimicked their original in more than just appearance. As Sebald observed, it wasn't only unfashionable adornments that were evicted or discarded. To make Brown's parks, the so-called Capability men physically tore apart the existing landscape. They dammed rivers to make new lakes, drained marshes to produce rolling lawns, slashed sight-lines through woods, moved full-grown trees to create the artful clumps that Brown's detractors disparaged as *puddings* and sculpted the earth itself into new forms.

Human beings were regarded as just as moveable as trees. The pristine vistas achieved at Harewood, Warwick Castle, Audley End, Bowood, Chatsworth and Richmond Park all required the removal of a hamlet or a village, complete with its forge and church, its inn and school, not to mention its occupants. Often this work was done piecemeal, over decades, so that for a time the as-yet-undemolished cottages stood uneasily alongside extravagant new pagodas or statue walks. More rarely, the old features were redeployed, as at Castle Howard, where the village street became a terraced walk to the Temple of the Four Winds. It's here that Jeremy Irons as Charles Ryder pushes Sebastian in his wheelchair in the television adaptation of *Brideshead Revisited*, a production so lavish and exquisite that it helped to reanimate the melancholy mythos of the country house decades after the war that had extinguished it.

If Ditchingham was misattributed, the landscape around our house was a Capability Brown created centuries after his death. Had Sebald made his journey by way of the Roman road today, he wouldn't have passed through the empty cornfields punctuated by dark plantations of trees described in *The Rings of Saturn*, but instead a perfect forgery of an eighteenth-century landscape garden. Brown came here in 1782 to *improve* Heveningham Hall, one of the many stately homes in the vicinity. It was probably around the time of his visit that our own house was being remade on Georgian lines from the rough materials of two Tudor cottages, its gardens surrounded by high new walls of brick and gault.

Brown drew up a plan for refashioning the park but it was never made on account of his death the following year. It was finally implemented more than two hundred years later by the property developer Jon Hunt, who made his fortune with the estate agency Foxtons. When he purchased Heveningham Hall in 1994, he discovered Brown's proposed improvements for the estate, the placement of eighty thousand trees precisely determined by his pen on a sheet of paper nine feet long. It was this design that Hunt brought into being, with the assistance of the landscape architect Kim Wilkie. In addition, he purchased five thousand acres of adjoining land and assorted buildings, branding the ensuing realm Wilderness Reserve, the word 'wilderness' deployed not in our current sense of natural or untouched, but in its eighteenth-century meaning of arcadia or pleasure ground.

All of the old estate buildings became holiday retreats: the gardener's cottage, the gamekeeper's cottage, the clock tower, the coach house, all the way up to the Palladian splendour of the big house itself. As with Brown's original landscapes, the settled, orderly appearance of Wilderness was a consequence of strenuous upheaval. Lakes were dug, trees planted, telegraph lines buried, unsightly roads and farm yards cleared away. The anachronistic attention to detail included ordaining that the estate's cleaning staff drive Morris Minor Travellers, which tooled continuously back and forth in front of our house. Even the sheep were stage-managed, angled against the sun, so they cast distorted knockoffs of themselves on the shining grass.

I crossed the estate most days. You had to in order to get anywhere. There was none of the dirt and damage that comes with age. It was all just-born, impermeable as a Gainsborough, depthless as Disneyland, the pilastered houses reduplicated in the silver mirrors of man-made lakes. There was a rumour circulating that autumn – the tree surgeon told me and I promptly passed it on – that Hunt had bought Cockfield Hall, a beautiful Jacobean house left empty for many years. Sure enough his bulldozers appeared, carving yet another lake from the old meadow.

That autumn I was reading John Barrell's revelatory book *The Idea of Landscape and the Sense of Place* alongside Sebald, and together they were opening my eyes to a network of power relations that isn't so much hidden in the landscape as expressed by way of it. The more I thought about it, the more it seemed that the expensive forgery of the eighteenth century in which we were now living was not an aberration or an anomaly. Instead, it was a perfect illustration of how the landscape garden was designed to articulate who possessed power and who lacked it, and how dismayingly little these relationships have changed in the past two hundred and fifty years. An entire ruling class ideology was embodied in this particular version of paradise, and it hadn't yet gone out of fashion or been redrawn.

Like *paradise* itself, the word 'landscape' is unexpectedly slippery. We think of it as referring directly to the land, but it originates instead with paint. It emerged from the Dutch in the sixteenth century as the technical term for a painting of a rural scene: a hunt in a dark forest, or harvesters lunching on bread and pears in a clearing of cut corn. An example of this usage can be found in a Suffolk will of 1648: 'I also give unto her Ladyship, the landskipp inamiled vpon gold which is in the Dutch cabinett in my closett.' From here, the word migrated by association to a rural prospect, one that you could stand and gaze at, a usage that according to the *Oxford English Dictionary* first occurred in 1645 in two places, one of which was Milton's *L'Allegro*. It wasn't until the mid-eighteenth century that *landscape* began to refer to a particular region, the landscape of Suffolk, say, or of the fens, by which time it was inexorably shaped by its involvement with concepts of visual value.

At around the same time, the land itself was being reshaped according to a painterly set of values. One of the most pervasive models for the eighteenth-century landscape garden was

Italian painting: Poussin, Salvator Rosa and in particular Claude Lorraine, a Frenchman based in Italy, whose expansive pastoral scenes became wildly fashionable in Georgian England decades after his death, though all but the richest buyers had to content themselves with a forgery or copy. Claude's paintings tend to repeat a similar composition: an awesomely well-balanced variegation of woods and waterways, hills and valleys, receding in irregular layers to the far horizon, as the land slips into evening. Whatever action the foreground holds – Greek gods in congress, the small dramas of farm or travel – is subsidiary in its appeal to the bright arena of the vanishing point, itself half-submerged beneath a milky wash of blue. Here and there the last light catches, beatifying a leaf, a branch, adding to the sense that the human drama is subordinate, almost inconsequential, compared to the majesty of the theatre in which it occurs.

What Claude's landscapes most powerfully convey is a sense of permanence and stability, of things occurring within the vault of the eternal. This remains true even when they include instances of violence or disorder, as in *The Battle on the Bridge*, which incorporates fleeing herders, a struggling crowd of soldiers and two people being flung or falling into a river without unbalancing the larger atmosphere of peace. They are profoundly apolitical, in a way that is itself political. The expansive prospect invites the viewer to take a godlike, dilated view, in which the human dramas seem natural and perpetual; a mood intensified by the time of day. Many of Claude's landscapes are depicted at the moment of illumination after the sun has set, when the air is briefly charged with golden particles and the land attains a mood of settled rightness, as if it was a model most ingeniously made of glass or spun sugar.

The popularity of these paintings helped to propel a new interest in the landscape, which might be coaxed by the trained eye into revealing itself along desirably Claudean lines. There was even a device to assist in the procedure: the Claude glass, a hinged box that opened to reveal a small dark convex oval of glass, behind which was set a mirror. The devotee turned their back on the landscape they wished to enjoy, and held up the device, which returned it in miniature and dimly, so that it appeared more like a painting of a landscape, imbued with a continuity and mystery that in the actual land could be in frustratingly short supply.

The Claude glass could make a landscape look literally picturesque, but with the help of Brown and his ilk, the ambitious landowner could go one step further, physically reshaping their acres along Claudean lines. The new landscape gardens created a simulacrum of the natural not by leaving nature to its own devices but by copying its idealised depiction in paintings. As Milton's admirer Horace Walpole observed in *On Modern Gardening*, written in a rush of inspiration on 2 August 1770: 'An open country is but a canvas on which a landscape might be designed.' (Pope, earlier and even more succinctly, to his friend Joseph Spence: 'All gardening is landscape-painting.') The parks might have looked more wild than the gardens that preceded them, but to use Walpole's own language, this was nature embellished, polished and chastened, as artificial in its creation as the stiffest knot-garden or heraldic bed.

The shift in taste towards the natural was, in short, composed of so many layers of forgery, replication and improvement as to make the Claude glass its ideal emblem. The Claude glass, or perhaps another invention of the same period: the ha-ha, which facilitated an even more alluring viewing possibility, that of the owner surveying the infinite prospects of their own landscape, which looks wide-open though it is in fact tightly cultivated and controlled.

It was my father who taught me about the ha-ha; my father, in fact, who carried out a subtle campaign of indoctrination into the pleasures of the landscape garden in particular, taking us to Petworth, to Longleat, to Burghley and especially to Stowe, with its temples and statues, its bridges, obelisks and shell-encrusted grottoes. I can't remember now on which green lawn he gave his lecture: the ditch with its double sides, to keep out deer and cattle while preserving from the house the illusion of unbroken parkland, stretching from terrace to horizon, hence the supposed exclamation of surprise, *ha ha!*, when the impediment to movement was discovered. My sister and I liked to scramble in them, peering out at placid cows, spinning round to view the house, perhaps engaged in a frenetic game we'd recently invented, concerning the outfits we might wear for a multitude of invented locations and pursuits.

In Walpole's essay, which is knowledgeable, gossipy, witty, not entirely accurate, and as mildly deranged as any exhibition of taste (a cousin, in short, to those tea towels that list the things the interior designer Nicky Haslam finds common), he claims that the invention of the ha-ha in 1730 facilitated the creation of the first truly English gardens.

No sooner was this simple enchantment made, than levelling, mowing and rolling followed. The contiguous ground
of the park without the sunk fence was to be harmonised
with the lawn within; and the garden in turn was to be set
free from its prim regularity, that it might assort with the
wilder country without. The sunk fence ascertained the
specific garden, but that it might not draw too obvious a line
of distinction between the neat and the rude, the contiguous
outlying parts came to be included in a kind of general
design: and when nature was taken into the plan, under
improvements, every step that was made pointed out new
beauties and inspired new ideas.

I read this paragraph several times before I realised what was
so odd about it. There is no active agent, no individual carrying
out the work of levelling, mowing, rolling. It just . . . happens,
passively, as if ordained. Walpole, the son of the first English
prime minister, Robert Walpole, is expounding here what came
to be known as the Whig view of history, in which time itself is
regarded as undergoing a continual process of improvement, in
which there are no wrong turns or costs to bear, but instead an
unwavering ascent toward the lovely summits of civilisation,
truth and beauty. Naturally. As if ordained.

The gardens of the eighteenth century are likewise a product
of this mindset, even if most don't illustrate it quite as emphatically as Stowe, with its Temples of Ancient Virtue and British
Worthies, its busts of Aristotle and Plato giving way to their
definitive heirs, Shakespeare and Milton. What confidence,

Athill had said, but wasn't it also a kind of confidence trick? To reshape the land in your own image, to reorder it so that you inhabit the centre and own the view. To fake nature so insidiously that even now those landscapes and the power relations they embody are mistaken for being just the way things are, natural, eternal, blandly reassuring, though what has actually taken place is the seizure of once common ground.

A harvest moon for Halloween, and with it the announcement of a new lockdown. I planted fritillary under the plum tree, along with the pink and white lady tulip, *Tulipa clusiana*. Another old gardening acquaintance of Mark came to visit. Matt and I walked a circuit together and he identified plants and gave me advice, telling me cheerfully what was beyond rehabilitation and what could be pruned into recovery. After he left I wrote pages of notes, trying to remember everything he'd said. I didn't know then how central Matt would be to the garden's regeneration.

A few days later it was the American election. When I woke at six it looked like Trump was in the lead. I'd spent the last five years immersed in politics, writing two books in succession about the shift towards fascism, digging around in the rubble of the twentieth century to try and find clues for a different kind of world. I never saw the garden as a place to escape reality, but I did think it was a way of training my eyes on a different aspect of it. The fragile aspect, it felt now. The perpetually destroyable.

First frost, each blade of grass furred with silver. I burned the toast, switching back and forth between screens. It was an immaculate morning. As the leaves fell there was a steady increase in light, leaking from the margins. It didn't help my mood. What kind of hell was about to be unleashed, for people, and for the vast, vulnerable world that people are so bent on making unliveable? I planted a gaura I'd lifted from my mother's garden, and then, as an antidote to staring at my phone, started to weed.

Slowly, the outside world receded, replaced by the ongoing epiphany of the everyday. A pheasant touched down and rocketed back up, jacketed in space-age emerald and bronze. Birds were clicking in the mulberry. I pulled out spent forget-me-not and dandelion. Somebody was shooting in the parkland behind our wall. Disputatious rooks were circling, calling out points of order, suggesting opportunities for integrative bargaining. I found more honey fungus in the beds, infiltrating an artichoke and a big stand of echinops, neither of which seemed troubled by its presence. The air was heavy with the scent of ripe figs. The garden was so beautiful, so liquid somehow, that it was hard to drag myself away. What arrogance to think that this is not reality too.

The lockdown would begin at midnight. While I was in the bath there were a few desultory fireworks and by the time I got out Wisconsin had been called. It dragged on like that for days. When the final state went blue I was back in the garden, raking leaves and bagging them for leaf mould. A moment later Biden's victory was declared and with it four years of constant anxiety dropped away. It was a strange, cartoonish feeling – the wicked witch melted, a red tie and an ugly wig lying damply on the floor – and I mistrusted it even as I luxuriated in the sense of a newly opened horizon.

One era does not replicate another, but I'm sure Trump's border wall had sharpened my feelings about the improver's park and what it stood for. When the language of exclusion is no longer coded but spoken unambiguously, one becomes increasingly alert to its disguised forms too. What I wanted to know was what it was like to be dispossessed by the work of improvement, which took so many forms and changed the landscape so drastically, and in this I was lucky, because someone was watching, sick at heart, and furthermore had stolen the time and scrounged the paper and even made the ink, from nut gall and green copper soaked in rainwater, with which to set down what he saw.

The first time I heard the name John Clare was in a poem. It was a short verse called 'Heard in a Violent Ward' by Theodore Roethke, about being institutionalised in Heaven; not such a bad fate if you could eat and swear with the poets. There was a list of possible candidates that ended: 'and that sweet man John Clare.' The phrase stuck with me, and so his sweetness and his madness were the first things I knew about the peasant poet of Northamptonshire, a man most unlike everyone else in the crowded halls of literature.

He was born, the only surviving twin, in 1793 in the fenland village of Helpston, a few miles from Burghley, one of the gardens most closely associated with Capability Brown. His father, Parker Clare, was an agricultural labourer whose rheumatism, caused by his work, made work increasingly impossible. The family's most reliable asset was a Golden Russet apple tree, the fruit of which might cover their rent in a good year, though failing harvests and the escalating price of food meant far more years were bad than good. Searching through his forebears, Clare could find nothing more exalted than 'Gardeners Parish Clerks and fiddlers.'

Clare himself was a dreamy, fanciful, ambitious, obstinate, prodigiously gifted boy, subject to fainting fits and often lost in his racing thoughts. In one account of his early life he reported himself a tenacious worker ('tho one of the weakest was stubborn and stomachful'), although in more honest moods he admitted his combined idleness and timidity made him alarmingly unfit for paid work. He went to school when he could be spared, his passionate interest in learning – 'improvement', he calls it, using the defining metaphor of the age – interrupted by the more pressing demands of the agricultural calendar (sceptical neighbours thought his habit of reading 'was for no other improvement then quallyfiing an idiot for a workhouse').

The fields were a site of duty and leisure alike. For nine months out of twelve he was under the sky, weeding or watching horses, bird-scaring or threshing alongside his father with a child-sized flail, though even full-grown he never topped a malnourished five foot. The delight of these jobs was to sneak away from them, in company or alone, dropping down amid the molehills to watch insects and birds, to hunt flowers and nests, to fish in streams, to read sixpenny romances like *Little*

Red Riding Hood or *Long Tom the Carrier*, or to practise his penmanship on the scraps of brown and blue paper that had once wrapped his mother's sugar and tea.

At twelve or thirteen he left school for good and was taken on as a plough boy by a neighbour. It was during this year of not unpleasant bound labour that his future sailed into view. A young weaver lent him a fragment of Thomson's long poem 'The Seasons', and what he read so compelled him that he begged the money from his father and walked five miles to Stamford to buy his own copy. It was a working day, and he was so greedy to read and so chary of being seen that he jumped the wall of Burghley Park and nestled there, 'and what with reading the book and beholding the beautys of artful nature in the park I got into a strain of descriptive rhyming on my journey home'. Later he set it down on paper: his first, as yet imperfect poem.

The 'artful nature' he'd admired was, of course, the handiwork of Capability Brown, who'd arrived at Burghley in 1754 and spent twenty-five years designing and refining the park. In the words of the Burghley guide, it was his genius 'which, brooding over the shapeless mass, educed out of a seeming wilderness, all the order and delicious harmony which now prevails.' Not long after jumping the wall, Clare returned to this harmonious space in the company of his father, in the hope of being taken on as a gardener's boy. He was hired as apprentice for three years, at a wage of seven shillings a week, and detailed to work in the kitchen gardens, taking lettuces and peas to the hall. It didn't last long. He was afraid of the head gardener, who was drunk and aggressive, and though he loved flowers, 'the continued sameness of a garden cloyed me'.

He couldn't write, or think, because writing and thinking came somehow directly from being outside, in the diverse landscape beyond Burghley's wrought-iron gates, the motley of heath, cultivated and fallow fields, woods and streams in which he'd grown up. The park might have been artful but the nature he prized was artless, wayward and untrammelled. Unlike his cultivated contemporaries, Clare didn't see nature as a shapeless mass and nor was he interested in discovering a lordly, sweeping, organising view. His preferred way of looking was belly-down in the grass, so that he could swim entranced into a microscopically detailed, teeming world. It was this immersed gaze that generated his poetry. He'd mutter the lines as he walked, or duck down to scribble them in the crown of his hat.

Here is the field song of Clare, fresh as the day it preserves. Nothing displaces anything else, everything is intermingled:

the varied colors in flat spreading fields checkerd with closes of different tinted grains like the colors in a map the copper tinted colors of clover in blossom the sun tand green of the ripening hay the lighter hues of wheat & barley intermixd with the sunny glare of the yellow charlock & the sunset imitation of the scarlet head aches with the blue corn bottles crowding their splendid colors in large sheets over the land & troubling the cornfields with destroying beauty.

Out it pours, detail advancing on detail, a continuous orison of praise.

After he left Burghley, Clare laboured at anything that came along. Militiaman at eighteen, short-lived, from which he returned with a prized copy of *Paradise Lost*, a poem that always made him cry at the end, when the angels with their flaming sword drove Adam and Eve out of Eden. In many ways it was the story of his life. The next year the family's cottage passed to a new landlord, who soon divided it and the garden between four households, upping the rent from thirty shillings to four guineas for what amounted to a single room.

In debt and deeper debt, the rent unpaid, his parents suffering the shame of going on the parish. Stints burning lime, that dirty white work, during which he fell in love, ill-timed since there was no money for marriage. Trying all the time to steal time in the fields, and trying too to send his poems into the world, not easy when what he was sending went so to speak in a filthy envelope sealed with shoemaker's wax. His origins were against him, though also a distinctive source of wealth, since no other poet making rhymes about woods or seasons had anything like Clare's dazzled clarity, the besotted precision of his eye.

Then, in 1820, a change in fortune. He married Patty and his poems in their grizzled envelope were published to the acclaim he'd dreamed of as a boy. *Poems Descriptive of Rural Life and Scenery* sold three thousand copies in a single year, far more than his contemporary Keats. He was given annuities by two local landowners, Lord Fitzwilliam of Milton Hall and the Marquess of Exeter, whose seat was Burghley. And now John

Clare was up and down to London, the celebrated poet and self-described country clown, in a borrowed black coat that fell to his feet and which he'd wear in even the hottest room, being ashamed of his clothes underneath. An inexhaustible appetite for beer, which was cheaper then than bread. Walking the city at night muttering *Oh Christ*, too frightened of kidnappers and goblins to pass down an unlit street. Still drunk when he got off the coach and returned to his wife, who might have thought, as Peter Quince did when he regarded the ass who had been his friend, *Bless thee, thou art translated*.

It isn't easy to tack between worlds, to weather the shifting winds of scorn and acclaim, to go up and down on a string, fate's plaything, and Clare, already high-strung, was sometimes unstrung altogether. Health failing, mood low, troubled by thoughts of death, pursued by blue devils, dreaming three nights of hell. Each successive book less successful than the last, the cottage filling up with sick babies, and Northamptonshire's peasant poet still labouring in the winter fields, marked out as different but not sprung free from his original station; a suspicious, unsettling figure who'd gone just far enough away to make it impossible to come home. 'I have been so long', he wrote, 'a lodger with difficulty and hope.'

All this is true, but in Clare's own estimation the hardest blow had fallen years before his poems were published, when in 1809 the act for the enclosure of Helpston was passed. Enclosure was another grand work of improvement, applied this time to agricultural land, which was likewise taken over and ruthlessly vacated for private profit. In the medieval system of open-field farming still practised in many places, including Helpston, a village was surrounded by two or three enormous fields, subdivided into many tenanted strips. These fields were farmed communally and on rotation, so that at any time some land was planted and some lay fallow. The uncultivated ground that lay beyond was categorised either as commons, which the farm labourers depended upon for grazing, foraging and firewood, or wastes: land resistant to farming, like fens, moors or marshes.

Enclosure was essentially a process of privatisation, a land grab legitimised by a flurry of new laws. The formerly open and shared fields were taken from the labourers who had tenanted them for centuries, redivided and fenced for their owners as private plots. The commons and wastes were brought into cultivation for the first time, an idea that Milton had once mooted, forcibly expelling the people who depended on them for their survival. In many areas of the country this shift had been going on piecemeal and by private agreement for centuries, but Clare was born at the height of the parliamentary enclosures, a highly organised and formal mechanism for reorganising land ownership and usage, in a place unusually unchanged from its medieval layout.

Parliamentary enclosure meant to modernise agriculture by making a more rational use of land, annihilating its original scattered dispensation and rearranging it to the greatest advantage of its owner. In this way it was analogous to emparkment. Indeed, enclosure sometimes provided the land for parks, or funded it by way of increased rents. Both were manifestations of a disregard for the existing landscape and the people who depended upon it, and both attempted to create a landscape of power, rigorously stripped of its human element. Between 1761, when George III was crowned, and the General Enclosure Act of 1845, around four thousand Enclosure Acts were passed, bringing more than five million acres of open fields, commons and wastes into private owner-ship and cultivation, a process of privatisation that has continued steadily to the present day.

The question of how grievously this affected the poorest agricultural labourers in Clare's own time remains a matter of debate among historians. There was a system of compensation for tenants and commoners who could prove their claims, and in some places it seems plausible that the loss of land rights was counterbalanced by an increase in available work, at least temporarily, while the physical reorganisation of enclosure was being carried out. But whether or not you agree with Marx that it was a form of robbery, in which 'the great landowners made a present to themselves of the people's land, which thus became their own private property', and with Orwell, that 'the land-grabbers . . . were quite frankly taking the heritage of their own countrymen, upon no sort of pretext except that they had the power to do so', the enclosure of the fields did make legible the painful fact that only those who owned it, as opposed to worked or knew or depended upon it, had unassailable rights over the land. It was, to quote a final voice of protest, the historian E. P. Thompson in *The Making of the English Working Class*: 'the culmination of a long secular process by which men's customary relations to the agrarian means of production were undermined.'

I agree with Thompson, for what it's worth, but this fundamentally economic account was not exactly what drove Clare's own sense of devastation. Like emparkment, enclosure was a rearrangement of the geographic as much as social order. The loss of the commons and the wastes, the draining of the fens, the levelling of hills, the cutting down of woods, the diverting of rivers, the stopping of streams, the division of fields, the putting up of fences and hedges, the building of roads and the closing of footpaths ('stopped up or destroyed as superfluous or unnecessary', as the Act for the Enclosure of Helpston put it): all these injured a different sort of relationship, an ecological continuum in which Clare felt himself both participant and loving witness.

In a famous passage about getting lost as a child on Emmonsales Heath, he wrote about walking along the furze 'until I got out of my knowledge when the very wild flowers and birds seemd to forget me.' His *knowledge* was another way of saying his familiar ground, the place he knew, but it also intimates, as John Barrell observes, that knowledge is itself a function of place, in which one's capacity to make sense of things, to generate understanding, is a product of being in some way rooted and at home, and that, even more strikingly, this sense of home is reciprocal: that one doesn't just know, but is known.

This particular way of thinking made sense to me. It was so much a part of what I was doing in the garden, learning it plant by plant, and in so doing tethering myself in place. Often at night, before I fell asleep, I went around it in my mind, deciding where I could plant orris root or fit a new pond, or simply passing from tree to tree, imagining the magnolia and the mulberry, out there in the not quite quiet dark. Each new plant I discovered, inspected and finally identified was like a stitch in a tapestry, populating the formless space. What had been generic green became specific, not just a recognisable cultivar or species, but an individual with its own idiosyncratic history and appearance. And as I did this work I changed too, my own history and mental landscape enlarging to encompass the garden I was making.

Clare's understanding of knowledge as reciprocal helps to explain why it was such a source of bitterness and despair for him to see the loved landscape torn up, Swordy Well, the Barrows, Langley Bush and Lee Close all destroyed, a single hazel left 'desolate' in what was once a grove of oak and ash. 'All my favourite places have met with misfortunes,' he wrote, and in poem after poem he returned to them, obsessively recreating their lost loveliness, mourning how they'd been stripped and despoiled. What he could see, and what the proponents of enclosure could not, was a delicate, intricate connection between things. The damage was relational, and so far more was lost than simply a way of making a living, let alone a place to look at, the aesthete's prized view.

If Swordy Well was ploughed up and turned into a quarry for stones for road-mending, then its loss reverberated through many other species. It was Clare who first articulated this, and his bees that 'flye round in feeble rings /& find no blossom bye' are the heralds of centuries of destruction in the name of improvement. Eden had been breached. It didn't help that in those hungry years the lack of work meant he often had to join the gangs who planted quickset hedges and put up fences, converting the land from an open mystery into a chequerboard of private units.

One of the things that most drew me to Clare was the strange role that gardening played in his predicament. It is a commonplace to state that his poems are a work of salvage, a way to continually remake a landscape that was imperilled or destroyed. In them he remembered the places he'd lost and the plants that grew there, preserving for posterity too their wonderful Northamptonshire names: *jiliflowers*, *ladslove*, *clipping pinks* and *blood walls*; the latter with its rich tawny streaks unmistakably the wallflowers I'd planted up against the potting shed. His descriptions are so exact that over four hundred wild and cottage garden flowers have been identified from his poems and letters.

But in the post-enclosure years Clare also worked physically to reconstruct Eden, or at least to gather up its shining remnants and make them safe, turning his own garden into a similar sort of ark as his poems. As he told his friend Rip, the painter Edward Villiers Rippingille, in a tempting letter designed to inspire a visit: 'at my back door I have a little garden which I cram with Flowers for I am foolishly fond of them & am now ambitiously striving to be a Florist tho with but little success at present.'

Florist then meant a person who grows flowers, not sells them, and as Clare's poetry declined in popularity he seemed to find a haven among them. His extensive library was full of gardening books, including a volume by the horticulturist John Abercrombie that he'd bought with his wages back when he was an apprentice boy at Burghley. The journal he started on Monday, 6 September 1824 (a very hot day, though forecast to rain) is a mixture of many things, a record of health and reading, a naturalist's notebook and a correspondence log, but woven right through it is Clare's account of his garden.

I loved reading that journal. It gave me a companionable feeling to find him noting down the day he lifted his hyacinth bulbs, or when the blackthorn came into flower, exactly the sort of thing I recorded in my own black A5 notebooks. October then was not so unlike October now, the Michaelmas daisy out, 'thick set with its little clustering stars of flowers', the horse chestnut losing its leaves, that 'litter in yellow heaps around the trunk.' He recorded discoveries on his walks and fantasised about making a compendium of poems called 'A Garden of Wild Flowers', or perhaps a botany written entirely in English, driven by his loathing for Linnean taxonomy, 'the hard nick-namy system of unuterable words'. By the end of the month the chrysanthemum had opened: claret, buff, paper-white, purple, rose-pink and bright yellow, though not the chocolate or coffee he'd hoped for.

Chrysanthemum were garden flowers, like the auricula that

Clare grew in a homemade theatre constructed in his shed, but his garden was also planted with wildings he dug up on his walks. Herb true-love from Oxney Wood, a yellow water lily to establish in a tub, *horse blob* (marsh marigold) in a damp corner, ferns of many kinds in a bed edged with box. In January he took earth from molehills for his flower beds and listed among the flowers now out the 'peeping' yellow aconite (a favourite Clare word) and a scarlet daisy. He lifted a sucker of barberry for his garden, and the next month a bush of ling or heather from Swordy Well, then furze bushes, his name for gorse, from Turnills Heath, and more of both from Ailsworth Heath, followed by iris, primrose and a *stoven* or stump of black alder.

Gardening in this way was not uncommon practice, but the specific places Clare went to and the plants that he retrieved tack very close to his enclosure elegies. His garden was like a living version of the 'midsummer cushions' the village children made each year, picking field flowers and sticking them into strips of turf as decoration, a custom he loved so much he named his final, unpublished book of poems for it. A garden is a more lasting place of preservation than a cushion of turf, and even though it couldn't hope to approximate the enormousness of what was lost, it did allow Clare to safeguard for a time the entrancement of wild forms.

On one botanising expedition he was mistaken as a poacher, writing furiously that these new 'joalers' made a 'prison' of the woods. Sometimes he liberated trees and planted them elsewhere. His enthusiasm for fern-hunting was overtaken by a mania for orchids (among his papers is a list of their names: cuckoo, bee, spider, lily leaved, spotted, military, female, Red Man, Green Man). As he hunted them he came across the enraging sight of men laying out the plan for a railway, just as the farm labourers do in *Middlemarch*. It would cut right through Round Oak Spring and Royces Woods, a new despoliation he recorded in despair, though unlike the yokels George Eliot finds so amusing he didn't attack the men with a hay fork.

The journal gutters out after a year, resuming briefly in 1828 with two distressed statements of ill health. By 1832, Clare was so unwell and troubled in mind that his worried friends and benefactors helped to secure him a cottage in Northborough, three miles from Helpston. There was a large garden and an orchard but the move seemed to undo him, to widen the breach, furthering an already agonising sense of loss and foreignness. The land was different in Northborough, out on the hard edge of the fens. He'd have no woods to walk in, no heath or furze-bush, so would be without the last remnants of a destruction he'd barely weathered. Sometimes he blamed enclosure for destroying his Eden, and sometimes this move, though he was also willing to consider wryly that it was a consequence of adulthood itself, the awakening into a cold and unadorned reality. 'Ah what a paradise begins with life & what a wilderness the knowledge of the world discloses,' he'd written years before, in an autobiographical fragment otherwise concerned with the rural festivals of his childhood. 'Surely the Garden of Eden was nothing more than our first parents entrance upon life & the loss of it their knowledge of the world.'

The world was going increasing ill with him. His Northborough letters almost split apart along lines of anxiety and apology. *I am not well. I am sorry. The declining state of my health.* He compares himself to a packhorse, nothing ahead but 'hard fare & bad weather'. He describes experiencing night-mares with his eyes open, though sometimes still there are flowers in his letters. He'd long been in the habit of exchanging desirable plants by post, particularly with his friend Joseph Henderson, the garden steward at Milton and his companion on fern-hunting expeditions. Rarities and dazzlements acquired in this way include a white peony, pink Brompton stocks, hepatica, polyanthus, Chilean glory vine and climbing snapdragon. Now, in 1836, he asks Henderson for woodbines and double blos-somed furze, the 'sultry' heath flower he most loved, though he was so weak by then that Henderson sent the package in the company of a man who could set it into the ground.

That furze might have been the last plant to go into Clare's garden. In June 1837, he went voluntarily to High Beech asylum in Epping Forest, where sometimes he believed he was the poet Byron or Lord Nelson or else a prize-fighter. He escaped in the summer of 1841, walking all the way from Essex to Northborough, 'foot-foundered and broken down'. He was at liberty that autumn but on 29 December 1841 he was certified insane and locked up again in Northampton General Lunatic Asylum. Though he was permitted to use the library, to write, to have hot baths, and to walk each day to Northampton; though he was treated with kindness and respect, he was not allowed to leave for even a night. He stayed in the asylum until the day he died, 20 May 1864, after which his body was finally sent by train, home again to Helpston.

There can be no refuge against this kind of loss. What was taken from Clare were the two things he prized most, his freedom and his home. In the asylum that he called a *'bad Place'*, a 'purgatoriall hell' and a 'Bastile', his letters grew shorter and more sporadic, sometimes losing their vowels altogether, but they still contained the free flowers of home, tokens of everything he now had to do without. In fact he spelled it out himself on a scrap of paper he gave to Dr Nesbitt, the asylum's medical superintendent: 'Where flowers are, God is, and I am free.'

In letter after letter to his son Charles, he asks after his family and then the flowers, as if both were kin. He describes how they once spoke to him 'incessantly' and adds – though there's a rip on the page – 'I had hopes I should have seen the Garden & Flowers [before] now.' In April 1849, he was still dreaming of his old plant-hunting expeditions. 'I very much want to get back & see after the garden & hunt in the Woods for yellow hyacinths Polyanthuses & blue Primroses as usual.' Six months later and with a touch of reproach: 'You never tell me my dear Boy when I am to come Home I have been here Nine Years or Nearly & want to come Home very much . . . how do you get on with the Flowers.'

Maybe he feared his memory was failing him. On a scrap of paper he listed nine plants, among them aconites and coltsfoot, March violets and Christmas rose, bare remnants of the luxuriance he'd once known. His final journal entry, written in pencil on 12 May 1850, when there were fourteen more years of life to serve, was so like the old nature notes you'd never guess at where he was. The plants still anchored him. 'Plumbs Pears & Apple Trees are in bloom & the Orchards are all blossoms'. It's a true testament to what plants can mean, how they can root and steady a person, as they have for me and for so many others – a testament too to the damage that is done when the relationship between people and the land is severed, deliberately and for the purpose of profit. Two months later, to Charles again, and so polite: 'I am still fond of Flowers.'

IV

THE SOVRAN PLANTER

I knew there was a frost because the lawn crunched when I went out to look at the moon. In the morning the garden was silver. The last dahlias had blackened and there were gleaming lines of spider web in the wisteria. As I walked into the pond garden something blue flashed beside the fig. I got down on my knees to look. It was a little iris, rising unexpectedly from a tatty nest of leaves in the lee of the wall. There were more flowers coming, rolled into tight sheaths, and so I picked it and took it into the house to identify. According to Wootten's *Plantsman's Handbook*, it was *Iris unguicularis*, the Algerian iris, which flowers in winter and is suited to the drought-stricken sand of East Anglia. It was a beautiful, rich blue, and I thought it might be 'Mary Barnard'. Indoors, it released a spicy scent that filled the sitting room, mixing pleasantly with the acridity of old wood smoke.

Even in December there was always at least an egg-cup of flowers for the house, or to take round to Pauline, my neighbour next door, ninety that year. The bunches of cosmos, colchicum and stripy dahlias had given way to the first delicate pre-Christmas snowdrops, followed by Christmas box, the sulphur-yellow bells of mahonia and the jumping jacks of winter jasmine. Gathering them in the morning was the inevitable prelude to more widespread clearance. I'd stoop to snip a mildewed aquilegia or skeletal opium poppy, and find myself surrounded hours later by piles of foliage, foxglove seeds in my hair.

Day by day, I was stripping the garden to its bones, weeding and cutting the dead growth away. Alkanet, chickweed, creeping buttercup, enchanter's nightshade, the rusty leaves of lady's mantle. The prunings went into the compost and the worst of the weeds were chucked into rubble sacks for the dump. Back and forth to the bins, bouncing the laden wheelbarrow up the steps into the empty garage and out the other side into the yard. This was another mossy, slightly mysterious space. It was high walled and silent as a well, paved with those grooved black bricks that look like chocolate bars and filled with a cobwebby clutter of ladders and rotting cherry logs, already housing nettles and ferns.

At the far end were two concrete chambers that must have been muckheaps when this was a working stable. The left hand bin contained the finished compost, a crumbly black heap that sometimes yielded old plant labels or rusty nails. The other side was Ian's domain, which he attended with the slow and occasionally maddening precision he brought to all manual tasks. He chopped the prunings into fastidious inch-long sections, layering them with mown grass and cardboard before tucking the not-yet steaming pile beneath a blue plastic tarp weighted down with bricks.

We had two projects that winter. The first was to strip most of the ivy from the north wall, as a prelude to introducing clematis and roses. Bees are passionate about ivy flowers and birds use it for shelter, but it had grown unchecked for a decade, forming overhanging battlements that blocked the rain and made the garden even darker and drier than it already was. Sometimes if you cut it with secateurs and tugged hard it peeled off in great satisfying mats of interlocking hairy brown roots as thick as my wrist. Mostly, though, it was ladder work, creeping along with a screwdriver and a hacksaw, prying it out of the mortar and tossing it onto the lawn.

The second job was to do something about the soil. In the main garden it was sandy and porous, incapable of holding water, and in the pond garden it was covered in a velveteen carpet of moss, punctuated here and there by the noses of new bulbs. The trunks of the hibiscus trees were mossy too. A poor omen, it meant the soil was so sour and compacted that there was no air or nourishment at the plants' roots. It needed enriching and aerating, but while I was still waiting to see what came up I couldn't dig it over. There wasn't enough compost to mulch the whole garden and at the end of November I'd bulk-ordered a tonne of manure.

Before it arrived, I had to clear the beds. I went out every day after breakfast, hood up if it rained, not stopping for anything, gulping down forgotten cups of coffee hours after they were made. The more I cut away, the more I revealed. Embryonic hellebore flowers, their pink and green petals tightly clenched. Japanese quince, the green buds breaking to reveal a flash of coral. I swept up leaves, yellow mulberry, golden hazel, stirring with them the rich, intoxicating scent of rotten figs. Only the beech hedge clung onto its copper, dropping elegant arched reflections into the black water of the pond. At night I planted in my head, imagining where I could put nerines or *Anemone coronaria*, to make pools of future colour.

On the last morning before the delivery it rained so hard the river burst its banks. The water in the air intensified all the greens, as if each droplet was a magnifying glass. The moss was practically vibrating. I sat on my heels, scraping and lifting each plush slab before tickling the soil, working my way from end to end with a hand fork. By the time I finished the rain had stopped and there were two half rainbows, not quite meeting. I went indoors and put up paperchains.

After that I ran aground. I had pleurisy again, though my diary records that on 8 December I went out anyway to spread manure on the pond beds, mulberry and rose borders. In the wake of this unwise act I was confined to barracks on doctor's orders, chest aching, through incessant days of rain. I watched the first snow from my window. While I slept, the builders arrived to repair the greenhouse, which they encased in a ply shell and then dismantled. Sometimes I could hear their radio, interspersed by hammer blows. *It's dark before 4*, I wrote. *Wisteria leaves everywhere.*

I emerged a few days before the winter solstice, just as the greenhouse was freed from its carapace. While the builders were packing away their kit, the youngest lad whitewashed the walls. The light ricocheted from it. I wasn't supposed to do heavy work and so I put the space back to rights with theatrical slowness, a single pot of pelargoniums at a time, though according to my diary I also dressed the Christmas tree and cleared more beds of leaves, raked the lawn, planted the peony 'Molly the Witch' and dug a new bed alongside the greenhouse. No wonder I kept complaining my chest hurt, or retired to bed each afternoon to nap and read detective novels.

By the solstice all the staging was in, the plants throwing exuberant midwinter shadows up the wall. It was the perfect place to spend the shortest day. I've always loved the winter solstice, just as I find the summer solstice weirdly disturbing. How can the light be diminishing, before summer's even got into gear? It feels as if everything's over before it's even started. In December the lack of alignment between day length and season is much more cheering. No matter how cold it gets there'll be a new infusion of light each day, inoculating against the damp bulk of winter.

It was lunchtime but I didn't want to go inside. I wheeled in an office chair and set up the potting bench, with pens and labels in a jam jar. I made a police line-up of rakes and spades and filled Mark's old bookcase with accumulated bottles of Tomorite and sachets of rose food, grease bands for fruit trees, hormone rooting powder, grass seed, reels of wire and balls of green string, seed pots and odd stones that had caught my fancy on beaches long ago. Bags of compost were stowed under one row of staging, with stacks of plastic plant pots under the other. As a final flourish, I hung a green soil sieve from a nail on the wall. Command centre. Operations room.

I'd always wanted a greenhouse. Even the name is delicious, suggesting two things at once: wild and domesticated, contained yet transparent, orderly but conducive to fecund growth. I'd been a connoisseur since childhood. I love a glasshouse of any size, from the Tropical House at Kew to the smallest allotment cold frame, especially when they contain ponds. There were dozens of photos from past pilgrimages on my phone: the cast-iron Victorian glasshouse at Belsay Hall, where you step into a beaded fug of jasmine and stephanotis. The shipshape cedar at East Ruston, layered with pots of begonias and plumbago in fairground scarlet and blue. My screensaver was a window at Kettles Yard, shelved so that the sun passes through pelargoniums and cacti, witch balls and glass dumps, casting florins of chlorophyllic light onto the floor. Then there were the ones I'd never seen and never would. Monty Don's tomatoes. Cecil Beaton's conservatory at Reddish, where a young David Hockney lounged in mismatched socks among the African violets, a smirking putto in a pink plaid suit.

There was a special pleasure to the greenhouse abandoned. In my photos from Somerleyton, the grapes went unpicked and ghostly nicotiana forced itself out of the glass vents in search of sun. At Shrubland Hall the cast-iron filigree of the conservatory had been reclaimed by creepers, like Sleeping Beauty's melancholy castle, the tables set for guests who hadn't come in years. I'd visited both these houses in the late summer, and they'd lodged in my mind: palaces of excess that had somehow become marooned in time.

Somerleyton is the better known, not least because it's the

first house Sebald visits in *The Rings of Saturn*, but it was Shrubland Hall that fascinated me. It was once one of the grandest houses in the county, a neoclassical palazzo set on a scarp high above Ipswich, its sprawling symmetry disrupted by a single colonnaded belvedere, designed so that guests could gaze out over what was once considered one of the finest Italianate gardens in England, a pleasure ground reached by a famous tumbling flight of one hundred and thirty-nine steps, modelled on those at the Villa d'Este.

Though the estate was privately owned, I'd discovered a public footpath that ran right through the park. The trees were some of the biggest I'd seen in Suffolk, giant sweet chestnuts and staghorn oaks that must have been standing when the house was first built. Signs of neglect were evident everywhere I looked. There were discarded fence panels and broken windows dumped against a wall. Peering through the gate, I realised it was the old kitchen garden, infested now with ragwort. Three workmen were hammering in stakes by the drive and since they didn't stop me I walked on up to the house. An older man appeared then at great speed, declaring that this was private land. I admired the house with the cravenness of the trespasser, and asked him what was on the other side. Forty acres of Italian gardens, he said promptly, venturing too that he lived in the groundsman's cottage, that he had been there twenty years, and had the biggest back garden in all of Suffolk, though it wasn't now in the state that he would like. Then he said again that I must return to the path, even though the house was empty and the owner rarely seen.

This encounter filled me with curiosity. When I got home I bought the only publication I could find: a Sotheby's catalogue from 2006, when the entire contents of the house were auctioned off over the course of three days, before the estate too was sold. The book was so big it couldn't fit through our letterbox. According to the introductory essay, Shrubland Hall was first built in 1770, while the rudiments of the gardens were laid out by Humphry Repton a decade later. It was subject to successive rounds of extravagant beautification by a single family of garden-makers, the Middletons; whose story, I realised as I read, almost exactly spanned the period from the writing of *Paradise Lost* to the death of John Clare.

I flicked on, encountering an extraordinary array of objects from every quarter of the globe. Marble busts, micromosaics of Rome, shoe buckles circa 1770, thirty-one old English pattern silver table spoons. A Meissen table service commissioned for Frederick the Great, painted with rococo sprays of flowers, a different arrangement for every plate. Majolica jugs, Aubusson tapestries, a Caravaggio, a mahogany spinet, Gibbon's six-volume *History of the Roman Empire* bound in calf-gilt. Fans, salt cellars, still lives, famille rose tea sets and famille verte baluster vases. I saw Lord Burghley's pale face amidst a run of portraits, black-clad and wary. Crewel work panels, embroidered with a humming network of insects on vines. A Japanned longcase clock, stuck forever at ten to two. It was an accumulation of centuries and continents, a robber baron's hoard, reassembled for the occasion into orderly categories before being permanently dispersed.

What that catalogue made clear was that the garden itself occupied a similar status. I prefer to think of gardens as dream-works, the result of intensely personal creative labour, but like the Meissen dinner service and the crewel work panel they're also status symbols and adornments, a way for money to announce its presence in a more comely or displaced form. But where does the money come from? That winter I became preoccupied with finding out how such an opulent garden as Shrubland was made, in economic as well as historical terms. What we might call the Middletons' multi-generational vanity project links together three continents, Europe, Africa and North America. It was a garden of empire, and like all such places it came at a cost so terrible its reverberations are still being felt.

The Middletons emerge out of the clutter of time in the Restoration, a few years after Charles II seized back the throne in 1660. While Milton was sheltering in his pretty garden in Chalfont St Giles, two brothers, Arthur and Edward, one a merchant and the other a mariner, left London for Barbados. The English planters who'd colonised the island in 1625 had converted it into an obscene factory for sugar cane, gambling on potentially enormous profits from this demanding and labour-intensive crop as a direct result of the forced labour of enslaved Africans. These men, women and children were brought by ship to Carlisle Bay from the West African coast, thousands every year, a constantly replenished labour force, since many died on the journey or were worked to death in the fields, mills and boiling houses of the plantations, where the cane was crushed, pounded, liquefied and heated to yield sugar.

In the seventeenth century more and more European countries were plundering Africa for slaves to develop their colonies in the Americas, using them to establish plantations for growing the tropical plants on which vast fortunes could be founded. *Nicotiana tabacum*, *Saccharum officinarum*, *Coffea arabica* and *canephora*. Tobacco, sugar, coffee. Not so long ago these products were luxuries for kings. Now they'd become the subject of frenzied acquisition by the middle classes. By the nineteenth century, they would tumble even further down the class ladder, serving as cheap stimulants for the working poor, an increasing number of whom had been dispossessed from the land by enclosure and set to work in the new factories of the industrial age, the first of which were sugar refineries.

This web of exploitation would eventually become so total in its reach that even a figure as obscure and parochial as John Clare was caught within it. His first poems were written on the scraps of blue paper that wrapped his mother's sugar and tea, shipped in from the colonies. The Clares were on the last rung of the rural economy, balanced precipitously above the workhouse, and yet in the hard years after enclosure their staple diet was bread and vegetables, washed down with weak tea. Sugar glows even more brightly than flowers in Clare's memoir of his childhood, a rare and covetable luxury, seen once a year at the Feast of the Cross: 'barly sugar candied lemon candied horehound and candied peppermint with swarms of colord sugar plumbs and tins of lollipops . . . ginger bread coaches and ginger bread milk maids.' He was forty the year the Abolition Act was finally passed, and so those sweet-meats set out on trestles at a village fair in rural Northamptonshire were the fruits of slavery, a system whose chains extended laceratingly right round the globe.

But this was very far into the future as concerns the Middletons. After the Restoration, the new king wanted to strengthen England's position among the European nations looting Africa, and in 1672 he granted an unrestricted trade monopoly to the Royal African Company in the traffic of slaves and other items, notably ivory and gold. It's within the Company's extensive archives that the exact nature of Arthur Middleton's activities in Barbados is revealed. He was what was known as an *interloper*, a slave-trader who operated illegally outside the monopoly. On 15 September 1675 and again on 16 June 1677, he was the subject of letters of complaint to the Royal African Company, that he was bringing enslaved people from Africa to Barbados in an unnamed ketch and in the ship *Alice*, both of which he part-owned.

This was the price of a stolen human life in Barbados that decade: £17 a head, or 2,400 lb of sugar. The planters complained that the Company drove up the cost, so that 'the poor planters will be forced to go to foreign plantations for a livelihood.' In return the Company complained that the planters did not buy enough of their wares, having 'so great a glut of negroes that they would hardly give them their victuals for their labour; and multitudes died upon the Company's hands.' People, that is, traded as commodities alongside elephant teeth, and with so little care that one in three did not survive three years.

Many of the planters did follow through with their threat and leave Barbados for America in the 1670s. They re-established themselves in England's newest colony of South Carolina, named for the new king and established by the Lords Proprietors, the eight friends rewarded for helping to restore him to the throne. Several of the Lords had shares in the Royal African Company, incentivising them to encourage slave labour in the province. In 1678 Edward Middleton joined the exodus, shifting operations to Charleston, followed a year later by his brother. He bought up thousands of acres of land grants in the Goose Creek area, establishing a plantation called the Oaks, and became prominent enough in this new community to serve as deputy to the Lords Proprietors.

At his death Edward's estates passed via his wife to his son, also called Arthur, who likewise served as a deputy to the Lords Proprietors, as well as president of the council and eventually governor of South Carolina. In addition to his political activities he too purchased thousands of acres of land and slaves, establishing a formidable network of plantations. The swampy terrain in Carolina was not conducive to the existing cash crops of sugar or tobacco, and the planters experimented with indigo and timber, as well as oranges, olives and silkworms, before settling on the rice known as Carolina Gold, which was sold as a cheap substitute for wheat in northern Europe. These planta-tions were dependent on the labour of enslaved Africans, who not only cleared the forests and worked the humid, muddy fields, but also provided specialised knowledge on how to construct and tend rice paddies, five hundred acres of which required the building of some sixty miles of dykes and ditches, levees, culverts and floodgates, every inch of which was dug by hand. Malaria was rife in the low country, along with typhoid, smallpox and cholera, making the summers deadly as well as unbearably hot.

By 1729, Arthur had built up such an abundance of land that he could afford to give his son William, recently returned from education in England, a plantation of his own, bordering on the Oaks and named Crowfield after an estate in Suffolk belonging to Arthur's aunt. The extent of his other holdings is evident from his will, which is full of specific bequests – '1630 acres where I now live', '100 acres I bought of the Lord Proprietors', '1300 acres at the head of Cooper river', '1500 acres on Wassamscue Swamp' – attesting to a lifetime of speculative accumulation. As the oldest son, William inherited estates in England, Barbados and Charleston, while the middle son Henry received a substantial plot in the Goose Creek area of South Carolina, including the original home plantation of the Oaks.

At his death in 1737 Arthur Middleton also owned at least a hundred and fifteen slaves. Since these stolen people were regarded by law as his property, he left them to his wife and sons too, listed unnamed after the disbursements of plate, linen, beds and glasses. While the two oldest boys followed in their father's footsteps as plantation owners, his youngest son Thomas became one of the region's most prominent slave-traders, importing people in chains from Gambia, Guinea and the Gold Coast to South Carolina, 3,700 in one single nine-year period.

Reading about this relentless accumulation, which was acquired by dint of refusing to acknowledge the freedom of other humans or to pay them for their labour, begs the question of what its purpose was. Money equals security, of course; education, status, a home. But after a certain point, what are those profits for? I mean this question seriously: what could you actually do with such a superfluity of wealth? Observing the gains made generation after generation in the Middletons' wills, I wonder what drove them, as I wonder what drives billionaires now in their relentless pursuit of capital. 'The impulse to mercantile accumulation': is that an answer, or does it just restate the question? To be rich, to build an unassailable paradise, a gilded kingdom outside time.

In *The Rings of Saturn* Sebald encounters a sugar-planter from the Netherlands in a Suffolk hotel, and it prompts him to observe the dominance of the country house as a reward. 'For long periods of time,' he writes, 'there was little scope for an ostentatious display of accumulated wealth, and consequently the enormous profits that accrued to the few families who grew and traded in sugar cane were largely lavished on the building, furnishing and maintenance of magnificent country residences and stately town houses.' Slave money, in short, crystallising into a new form, growing by the generation ever more refined. In South Carolina, what was true elsewhere of sugar was also true of rice, a trade so buoyant that in the eighteenth century it made Charleston one of the richest cities on earth.

The Middletons did exactly as Sebald proposed. As soon as he moved to Crowfield, William Middleton built a lavish Palladian mansion, surrounded by a sophisticated garden. While a handful of decorative gardens predate it, Crowfield was the first landscape garden in America, made to the height of European fashion and funded in the most repellent way. Its ambitions are best grasped now by way of a letter written by an enchanted visitor, herself a plantation owner, in 1742. She described an elegant house reduplicated in a mirror pool. At the back there was 'a wide walk a thousand feet long; each side of which nearest the house is a grass plat ornamented in a Serpentine manner with Flowers.' On the right, there was a thicket of live oaks known as a *bosquet*, and on the left a sunken bowling green, bordered by a double row of laurels and catalpas. 'My letter', the visitor continued, 'will be of unreasonable length if I don't pass over the Mounts, wilderness, etc, and come to the bottom of this charming spott, where is a large fish pond with a mount rising out of the middle the top of which is level with the Dwelling house, and upon it is a Roman temple.'

Not to be outdone, William's brother Henry also created a lavish garden at yet another plantation. He acquired Middleton Place by way of an advantageous marriage in 1741, which made him one of the richest men in the province. If Crowfield was sophisticated and at the forefront of fashion, Middleton Place was and remains an almost unbelievable feat of engineering, not so much a garden as a deft resculpting of the land, formed into terraces, canals and pools, culminating in a butterfly lake that juts out into the Ashley River. Like Crowfield, it utilised the techniques of the rice-planter, the subtle management of earth and water, deploying the profits of the nightmarish work there to fashion its twin, a landscape as purely consecrated to idleness and pleasure as the original was to enforced labour. It is tethered to its origin place, arising out of and encircled by the rice plantation, and at the same time it floats free, elutriated, an earthly paradise founded on exclusion and exploitation, which conveys to even the most casual observer the owner's absolute mastery over resources.

Not that Henry Middleton built it, of course. Like the plantation itself, the house and garden were constructed by slave labour. It's been estimated that the earth and water works needed to make the gardens at Middleton Place took one hundred slaves a decade to complete. No one recorded their specific identities, but the volunteer historians at Middleton Place have recovered from family documents the slave names of hundreds of people owned by Henry Middleton and his descendants, among them Cymon, Mercury, Scipio, Hagar, Doll, Moll, Nanny, Glasgow, Dido, Prince, Bob and Winter, who worked on the gardens, Adam, Cupid, Caesar, Answer, Hercules, who went to the Indian Nation, and Kouli-Kan the cooper, who ran away.

Christmas came and went. I spent the last days of December making my seed order. *Tagetes patula* 'Burning Embers', *Dianthus caryophyllus* 'Chabaud', *Papaver somniferum*, *Lathyrus odoratus* 'Spencer mix' and 'Prince Edward of York'. Marigold, pinks, poppies, the old-fashioned, heavy-scented sweet peas. I threw in some nasturtiums too, and a variety of cosmos: 'Rubenza', 'Dazzler', 'Daydream'. We built a leaf-mould bin in a neglected corner, discovering a broken chimney behind the yew. There was a succession of hard frosts and on the last morning of the year I found a single frozen hellebore, like a sugar flower.

The word 'planter' was looping through my head. It sounds so benign, in its original sense of a person who plants bulbs or seeds in the ground. The more sinister second meaning was acquired late in the sixteenth century: the founder of a colony. By 1619, the colonising work of planters in Virginia had given rise to a third and more specific meaning: the owner of a plantation. By then the word 'plantation' too had undergone a migration, from the act of planting seeds to the act of establishing a community of people on what the *Oxford English Dictionary* emphatically calls conquered or dominated country. A plantation might contain gooseberries and raspberries grown from suckers or cuttings, but then again it might be people, suckering from the homeland and re-establishing overseas, in someone else's home. Indeed, the ship by which the original Arthur Middleton travelled from the colony of Barbados to that of Carolina was named the *Plantation*.

Does language follow events, or do ideas already formulated in language shape and drive the way things happen? The story of Eden is lodged right at the heart of colonial endeavours. The desire to seize it, to claim and possess a paradise of new resources, was among the major drivers of colonisation.

But Eden also served as a justification, a God-given excuse note for the brutal work. In the seventeenth century, arguments for colonial expansion regularly drew on Genesis, and God's injunction to man to subdue and have dominion over all creation; an attitude, I might add, that is directly responsible for the perilous state of our planet now.

As John Smith, the Jacobean explorer and one of the first governors of Virginia, wrote in his last book, *Advertisements for the unexperienced planters of New-England, or anywhere, or, The path-way to experience to erect a plantation*: 'Now the reasons for plantations are many: Adam and Eve did first begin this innocent worke to plant the earth to remaine to posterity, but not without labour, trouble, and industry.' The establishment of colonies was seen as continuing their divinely appointed labour, expanding into what Smith disingenuously describes as a great excess of land, more than either 'the people of Christendom' or 'the natives of those Countries' could possibly tend or care for. Here, Smith claimed, the gospel could be planted alongside seeds, so that spiritual and economic profit could be cultivated together.

Yet this self-serving story came accompanied by its own anxious shadow, the possibility that land was being stolen and human inhabitants butchered or dispossessed. Rumours of savagery seeped back to England, stirring disquiet about the planters' work. Was Eden in fact being invaded, a pristine wilderness of boundless fertility sacked and spoiled? Born a year after the colony at Virginia was founded, Milton had been absorbing these contradictory narratives all his life, particularly during his years at the centre of the Interregnum government, as Secretary for Foreign Tongues. One of the many ways of reading *Paradise Lost* is to see it as emerging out of a century of unprecedented colonial expansion, its vision of an immaculate wild garden inflected by reports that were carried home from newly conquered lands. We say such things are in the air, and traces of colonial activities in the New World of America and the West Indies can be found all over Milton's Eden, like items washed ashore after a shipwreck.

On the most superficial level, Milton's similes draw continually on imagery of these newly encountered lands, just as they draw on classical scenes and locations. After his arduous journey to the 'new World' of Eden, Satan is met by gusts of sweet air; in the same way, Milton writes, that sailors travelling round the Cape of Good Hope are beguiled by perfumes emanating from the shore. The land he sees is a marvel of fecundity and abundance, described in language that borrows closely and sometimes word for word from the paradisiacal accounts of America sent home by astounded explorers. Even the trees are the same species, giant cedar, pine and fir. Adam and Eve are explicitly compared to 'th' *American* so girt / With featherd Cincture' who was encountered by Columbus, 'wilde / Among the Trees'. Meanwhile God is described as 'the Sovran planter', who established Eden as 'th' addition of his Empire'. Eden is literally a colony of Heaven, though in a different way so too is Hell.

Satan is exactly the kind of rebellious and undesirable reject displaced onto the colonies. He is expelled from Heaven by way of a succession of fairly unadorned intestinal metaphors, until with great groaning he and the other rebel angels are shat into the sewage system of Hell, an image much in circulation in seventeenth-century arguments for colonies as a place to house the unwanted human refuse of the home nation. At the same time, he's a coloniser himself, a 'great adventurer' plotting the invasion of Eden, who betrays and forces into servitude its naked and innocent native inhabitants, and invites his own relatives, Sin and Death, to take up residence in their stead. Then too Adam and Eve are indentured servants, who God will 'plant' to labour the new land, diligently pruning back its wilderness, who if they work hard and obediently will one day receive the reward of Heaven, the original and pristine homeland.

This succession of possibilities derives from the illuminating *Milton's Imperial Epic* by J. Martin Evans, which finds evidence of multiple overlapping kinds of colonial narrative at work in the poem, whose mutual incompatibility adds to its abiding power, its refusal to resolve as pure allegory of colonial misdeeds or, on the other hand, as colonial propaganda and boosterism. In short, Eden is a place of infinite abundance and possibility, where someone else owns the land, where adventurers plunder and dispossess, where labour is ordained and never-ending, where there are bosses and enforcers, where eviction is an omnipresent threat and at the last a tragic reality. It sounds a lot like Earth.

What I find so interesting about Evans's account is that it suggests the impossibility of Eden as an apolitical, ahistorical space, in that every story told about it also incorporates it into a political framework. The fact that sometimes these different Edens wrestle against each other is not the same as thinking the garden of paradise exists outside time and history. Like many people who benefitted from colonisation, the Middletons might have believed themselves to be so many well-dressed Adams and Eves, virtuously inducing the land to be ever more fruitful, though not of course with their own sweat and toil. They certainly thought they had paternalistic relationships with the people whose lives they stole, persuading themselves that those they'd enslaved were at once subordinates to be commanded and somehow also grateful for that command. That this was very far from being the case emerges in the second half of their story, in which the grotesque way their Eden-making was funded generated yet another Miltonic occurrence: civil war.

In 1754, William Middleton inherited the English Crowfield, a chance bequest that brought the family's enormous profits from slavery back to England for the first time. The move initiated a new phase for the family, in which the wealth they'd accumulated in America was further purified back home, in part by way of their obsessive garden-making, a process that helped to lever them into the very highest levels of society. To make a modern analogy, they used gardens to cleanse and frame their reputation, just as the Sackler family used art to elevate themselves, to rise above the degraded and exploitative sources of their wealth.

The estate at Crowfield in Suffolk had originally belonged to Edward Middleton's brother-in-law, who likewise had Barbadian interests, owning and investing in sugar-cane plantations. By the time the inheritance reached him, William was done with America, perhaps because within a handful of years he'd buried two wives, his father and four small children (the high death rate in South Carolina helps to explain how individual members of the plantocracy became so wealthy, since widowers, or more commonly widows, could end up absorbing two, three, even four estates).

There's no record of any garden-making at the Suffolk Crowfield, a house that has long since been destroyed, and the next stage in the Middletons' work of beautification was led by William's oldest son, also called William. In 1775, just as the War of Independence was beginning in America, this latter William made one of those advantageous marriages at which the Middletons were so skilled. His wife was Harriot Acton, the daughter of a Suffolk landowner, who furnished her with a substantial dowry of £8,000. The happy couple can be seen a decade later in a naive portrait by the Suffolk miniaturist John Smart, now flanked by their three children.

They are clearly outside, against a tobacco-coloured sky, sheltering beneath a scarlet canopy with dangling tassels. A tent? A bed? The girls are dressed in white with blue sashes and matching picture hats, their hair in formal ringlets. The boy, not much more than a babe in arms, sits on his mother's lap, reaching for her hair, constructed after the fashion of the time into elaborate grey coils. William Middleton, Harriot Acton and their oldest daughter are all staring fixedly into the lower right-hand corner, and something of the blankness of their expressions, their little dark eyes in slabs of pale faces, irresistibly recalls plaice, dressed up in collars and ruffs, contemplating nothing. Only the standing girl, Louisa, looks at the viewer, her hands raised strangely, in a *presto* gesture, as if she were the conjuror responsible for this piscine metamorphosis. At the very back there's a little cottage set amidst a clump of trees. Harriot's father had been painted by Gainsborough, but there's nothing here to match the boastful expanse of land in his gentry portraits like *Mr and Mrs Andrews*. The family might be travelling across a desert in their tent. Perhaps this is what they're gazing at so avidly: their true home.

They didn't have long to wait. After his father died in 1785, William purchased Shrubland Hall, built fifteen years earlier by the Palladian architect James Paine. The hall was at the time a more austere and restrained building than the palazzo that now dominates the landscape, though still the newest and grandest house in the vicinity. It was situated in an open park, containing even then the sweet chestnut trees I'd admired. 'Thus,' says the Sotheby's catalogue admiringly, 'the wealth accrued by the Middletons over the previous century in America found a physical expression in their native Suffolk' – though in fact the family was originally from Twickenham and their wealth was accrued by the most grotesque means.

It was to this park that William now turned his attention and his wallet. He was using the money accumulated in America to reinvent himself as an English gentleman in a country seat, and a garden was central to his ambitions. Like Mr Rushworth in Jane Austen's *Mansfield Park*, a novel fascinated by the possibilities and perils of such reinventions, he was unsatisfied by his new domain and itched for improvement. For the fashionable inhabitants of Mansfield Park there could be only one possible answer to this dilemma: the landscape designer Humphry Repton, who could contrive the most impressive alterations to the prospect of a house. 'I think I had better have him at once,' says Mr Rushworth, adding a little anxiously that Repton's terms are five guineas a day. Not everyone at the table is so eager. The heroine, Fanny, rues the avenue of ancient oaks that will be chopped down, while her conservative cousin Edmund admits he would rather change a place slowly and according to his own taste than submit to the totalising vision of the improver.

By the time Austen began *Mansfield Park* in 1811, Repton was the undoubted heir to Capability Brown, the magician who could clothe the tired parks of the gentry in what Austen calls 'modern dress', slashing down trees, installing shrubberies for ladies to wander arm in arm, and opening up such appealing vistas that an old house could feel bracingly renewed. But when he arrived to view Shrubland Hall, on a wet July day in 1789, he was right at the beginning of his career. It was only the third of the four hundred gardens he designed. Even as a novitiate, Repton had established his famous working method of the Red Book, though in the case of Shrubland its bindings are actually a much scratched green. It contains his before-and-after vision for the park, provided in 'minute detail', so that it could be implemented gradually, as time and budget permitted, without losing the sense of a uniform plan.

Many of Repton's changes were made by the Middletons – no doubt with all the dirt and confusion, the upsetting of gravel paths and rustic seats, that Austen's Miss Crawford predicts – along with ideas by a second designer, William Wood, whose tastes inclined even more to the picturesque. The park was extended and given a grander approach, as well as new walks, made interesting by the introduction of less common types of tree, including Lombardy poplar, larch, spruce and lime. As it flourished around him, a foaming green bower, William Middleton became an MP and then a baronet, taking on the name William Fowle Middleton; motto *Regardez mon droit*, *Respect my right*. By improving the garden, he had improved himself.

Money flowed inexorably in his direction. In 1801 his brother Henry won £20,000 on the national lottery, which he shared with William, and in 1811 Henry died childless and unmarried, leaving William as his heir. These events sound neutral, as if fortune happened to be smiling on and on, and yet the upsurge in wealth and power was still fed by what had by now been completely excised from view: the fact of slavery. We might think again of *Mansfield Park*, itself a sugar-planter's mansion, where Fanny's attempt to raise the subject of the slave trade with her uncle, newly returned from his estates in Antigua, is met by an impenetrable cushion of 'dead silence' from the rest of the family, the English defence against any painful or exposing subject.

Let's slice through that silence. When the original William Middleton left America, he didn't relinquish all his interests there. He still owned more than nine thousand acres in Carolina and Georgia, all run on slave labour. These were left to his three younger sons, John, Thomas and Henry, meaning that William Fowle Middleton did not initially receive them. But his brothers predeceased him, and so by various convoluted routes he received ongoing as well as historic wealth from slavery. The family papers, which were stored at Shrubland until its sale, after which they were moved to the Suffolk County Records Office in Ipswich, contain many letters in fading brown ink referring to these American properties and their shifting ownership. 1778, purchase of land by Thomas Middleton on River Ohio. 1783, repurchase of the original Crowfield in South Carolina by John Middleton, who died a year later. 1786, concerning Middleton property and affairs in South Carolina and England. 1792, concerning Henry Middleton's rice crop. 1796, concerning land in Georgia. 1798, concerning excellence of cotton crop.

In 1799 Henry Middleton wrote a letter from Bath, complaining of his health and expressing a wish to dispose of 'his property and his negroes' at one of his five plantations, Enshaws. This population, he was informed, comprised '104 workers, 13 half task hands, 48 children plus infirm and disabled', estimated at a value of £2,500 to £3,000. In 1806, he returned the appraisal of their value unsigned, callously suggesting that a public sale, which almost always entailed family separation, would improve the price. On it goes, an income stream welling up in the Carolina swamps, enforced labour and human souls converted into capital, generating a prodigious paper trail of chancery suits, mortgages, land transfers, rates of commission and bills of exchange, all of it deriving from the stolen work of stolen people.

When William Fowle Middleton died in 1830, the deaths of his brothers meant he had come into possession of much of their land. Among the many things he left to his eldest son, the absurdly named William Fowle Fowle Middleton, were Shrubland Hall, his baronetcy, and a bond on a 7,400 acre plantation in Carolina, along with 183 slaves. This estate was valued at £21,250, bearing interest at the rate of 7% per annum. Though the younger William spent considerable effort trying to sell either the bond or the land itself, he received to the end of his life an income from slave plantations owned, leased out or mortgaged in America.

How was it spent? What purpose did he put it to? More gardens: a flowering pathway to influence and power. Like his father, William had married well, to Lady Anne Cust, the daughter of the fourth Baron Brownlow. The couple punctuated their time in Suffolk with long stints in Italy, pursuing the nineteenth-century version of the Grand Tour and buying up treasure with which to decorate every last hallway and terrace of their estate. The Caravaggio of soldiers gambling in the Sotheby's sale; that was theirs, along with a great medley of porcelain, glass, silver, medieval chairs, animal skin rugs, though it must have been William's father who bought the Sevres china at the sale of the late Queen Charlotte, whose personal possessions were auctioned off by her perpetually hard-up son, the Prince Regent. Listing these items, the writer of Sotheby's catalogue, surely no stranger to aristocratic excess, was moved to comment: 'The sheer number of objects and their diversity is almost bewildering.'

This orgy of nesting, which amounts to a kind of horror vacui, was accompanied by near-constant work remodelling the house and garden in which their new possessions were to reside. It might be thought that there was little more to do at Shrubland, but William Fowle Middleton's death had initiated a final salvo of improvements, an influx of money and effort that saw the house and garden transformed from a luxurious gentleman's residence into an idiosyncratic and imposing palace on a hill, an opulent recreation of Italy amidst the Suffolk fields. Three successive architects, James Gandy-Deering, Alexander Roos and Sir Charles Barry, the architect of the Palace of Westminster, were brought in to redesign the hall, effectively entombing the original in such an encrustation of new rooms, vestibules, entrance halls and corridors that it recalls the home-making of the caddis fly. Roos designed the conservatory, while Barry was responsible for the asymmetrical tower, as well as a portico with arcaded wings, containing sculpture galleries ('far more lavish,' says Pevsner, 'than the rather austere exterior would suggest').

The old landscape park created by William's father would hardly do as a foil to all this magnificence, and in the 1840s Barry designed the great staircase, modelled on that of the Villa d'Este at Tivoli. If you descended it, entering through a small stone pavilion called the Temple of the Winds, and perhaps pausing on one of many landings, edged by box and flanked by urns, to gaze out at the brilliant spectacle below, you would see a most elaborate network of gardens, covering one hundred acres, all of them apparently designed by Lady Middleton. Which would you choose today: the Panel Garden, the Loggia, the Green Terrace, the French Garden, the Fountain Garden, the Poplar Garden, the Rosary? On it goes, until your head is reeling. The Hanging Basket Garden, the Box Terrace, the Dial Garden, the Witches' Circle, the Maze, or perhaps the Swiss Garden, at the centre of which was the first Swiss Cottage in the country, containing relics of Nelson and Napoleon.

Each discrete space was further embellished with such an array of marble statues and stone figures that it was as if the Medusa had gone on a rampage through the laurel walkways. With Victoria on the throne, the fashion had swung from the naturalistic or picturesque landscapes of Brown and Repton to the monumentally artificial, and Shrubland epitomised the new taste. It was emphatically a garden of empire, a machine for generating lucrative new connections. William was sent seeds from Constantinople, accompanied by comments about the Crimean War, and plants shipped from Brazil, selected with personal advice from Sir William Hooker, the first director of Kew, who not only took the time to write but also commented that the Duchess of Cambridge and Princess Mary were charmed by the Shrubland gardens. But that was what the garden had been made for, surely: to beguile the crown, to insinuate the Middletons into the highest echelons of society. In 1851, Prince Albert came to dinner at Shrubland, planting a

Liboudros pine tree to commemorate his visit. In return, William was invited to join a shoot by the Palace.

I part from Princess Mary here. I viscerally dislike Shrubland as it appears in Adveno Brooke's *Gardens of England*, from which the descriptions of its layout are drawn. Lady Middleton was a pioneer of the Victorian fashion for bedding out, using half-hardy, so-called 'exotic' perennials like white petunias from Brazil and yellow calceolaria from Chile to create geometric patterns of strident colour. Her contribution was to invent the 'ribbon band', a narrow strip of colour at the front of beds, as well as the prac-tice of 'shading': cramming hundreds, say, of different red geraniums to create a migraine-inducing throb of scarlet. Under the head-gardener Donald Beaton and his successor Mr Foggo, eighty thousand annuals were required for these displays every year, in addition to the vast number raised from seed.

I saw the final spasm of this gruesome style in the Beach House Park in Worthing, my grandfather's favourite place for a morning stroll. In the 1980s, the beds were still rigidly arranged medallions of hyper-real, chemical colour, each flower stiffly identical to its neighbour. Roses were segregated by tone and every tulip was a straight-spined sergeant at arms. Individuality was not permitted, and though I loathed it at the time and all my own gardening is set against it, I see it now as a municipal monument to the wartime virtues of collective effort and self-abnegation, which is not at all what Shrubland was attempting to convey.

January. My diary filled up with lists: prune apples, cut back honeysuckle, turn compost. I ordered plugs of flowers for the spring, nothing like the Shrubland kind. Soft white foxgloves to plant under the hornbeams, nicotiana for Ian and three dahlias, 'Wizard of Oz', 'Nuit d'Eté', a clear deep red, and 'Ambition', which is every gardener's curse. The work was my work and I didn't mind how hard it was. Though I often felt overwhelmed or out of my depth, I also knew that to be able to make something on your own terms is among the greatest luxuries, though it should also be a universal right.

We watched Boris Johnson on the television, announcing the start of another new lockdown. He looked old. Every day felt the same, suspended in amber, and yet history was rushing past us. Trump's supporters invaded the Capitol, and a Democrat won the run-offs in Georgia. I cut the first fragrant twig of daphne. There was nowhere to go, no end to the trouble we were in. In the morning I transplanted primroses while Ian stripped more ivy, until the walls were ready for what I described hopefully in my diary as a new era of exuberance, though it felt more like the end of everything. Often I went out at night to check the stars, Pleiades, Orion, Cassiopeia wheeling on her burning throne. In to look at the constellation maps and back again to regard the universe riding above the magnolia, leaving a chain of prints behind me on the grass.

An end of sorts was coming for the Middletons, too. The 1833 Abolition Act had passed while Lady Anne was planting out her shaded borders. Her brother, General Sir Edward Cust, was awarded £5,029 7s 8d compensation under the 1837 Slave

Compensation Act, part of the obscene £20 million paid out by the British government to slave-owners. Cust's payment was due because he was the heir to his mother-in-law's estate. Mrs L. W. Boode had owned a sugar plantation in British Guiana where 201 people were enslaved. Not one of them received any compensation for what had been taken from them.

The Middletons didn't receive compensation either, since they'd long since sold off their West Indian holdings. Abolition only applied to certain territories in the British Empire, and because the Middleton estates were in America they still profited from slavery after 1833. There is plentiful evidence in the family papers accounting for what this meant financially, pound by pound, and almost nothing about what it cost the people concerned, beyond those bare lists of names, carefully annotated with their price.

The dead silence Fanny describes in *Mansfield Park* has been the subject of considerable debate. It's possible that she is simply referring to her cousins' lack of interest in any subject besides fashion or romance. But then *dead silence* is such a striking phrase. It floats beyond itself; it captures, even if unintentionally, what even now cushions and muffles the subject Fanny tried to raise. One of the operations by which capitalism perpetuates itself is displacement, the determined and absolute separation of the product from the site of production, so that when we buy petrol or peat from a garden centre or even a chocolate bar, when we turn on a light switch or a tap, when we flush a toilet or purchase a sofa from Ikea, we are persuaded to believe that these things have arisen spontaneously, naturally, as a magical response to necessity or desire, while their real and often destructive origins and after-effects are determinedly concealed, as if banished behind a screen. This displacing operation is so powerful that to insist on showing the connection between product and process induces a kind of incomprehension, even fury. It's somehow wounding to the consumer to have to connect the can of Coke, a sleek and gleaming apparition, with water shortages and pollution in Uttar Pradesh, Kerala and Rajasthan.

A single member of the Middleton family refused this code of silence, attempting, however clumsily, to show the source from out of which all those beautiful houses, gardens, paintings kept emerging. Funnily enough, her name was Fanny too. She was the English actress Frances Kemble, renowned in Victorian London for her performances of Shakespeare, the eldest daughter of the actor-manager Harry Kemble, of the famous theatrical family. In 1834 Fanny married a slave-owner from Georgia, Pierce Mease Butler. Pierce was the great-grandson of the slave-trader Thomas Middleton, and so William Fowle Fowle Middleton's second cousin once removed.

Thomas's daughter, Mary Middleton (the first cousin of the first William to own Shrubland), had married Pierce Butler, an Irishman who served as South Carolina's adjutant-general in the War of Independence. Pierce was one of the Founding Fathers who signed the Declaration of Independence. He was also one of the largest slave-holders in America, responsible for introducing the ugly Fugitive Slave Clause into the Constitution, which made it far harder for slaves to escape into the free states in the north. He had disinherited his son, and so his plantations, situated on two islands in Georgia, were divided between two grandsons, one of whom was Pierce Mease Butler.

Fanny Kemble was opposed to slavery, though not so opposed as to refuse to marry a slave-owner. Her new husband persuaded her to visit Butler Island, promising that her opinions would change after experiencing the reality of plantation life. She arrived on the island in the winter of 1838 and kept a diary of her experiences, which she later published as *Journal of a Residence on a Georgian Plantation in 1838–1839*. This document represents a tiny breach in an otherwise totally dead silence: a rare acknowledgement by a member of the Middleton family – indeed, by any member of the planter class – of the 'prison-house' on which their fortunes were founded.

Her book is an eye-witness account of daily life on an ordinary plantation. The lack of food and brutal overwork were ordinary, as was the inadequacy of clothing, bedding and housing, the constant beatings, the brandings, the prevalence of rape by white overseers, the forcing of field labour in pregnant women and new mothers, the absence of healthcare, the miscarriages, the injuries, the disabilities, the deaths; above all the constant threat of family separation, as a punishment or simply because an owner decided to sell off the people he regarded as his property. The enslaved people told her their stories, and for the most part she believed them, against the accounts of her own husband. She saw the absolute deforming ugliness of white supremacy, even if her account is continually distorted by her own endemic racism.

Journal of a Residence is a difficult book. For one thing, she is a white woman riding around a plantation on a horse, the epitome of privilege, Lady Bountiful with a pen. Her account is distorted by the deep racism of her period and class, by sentimentality and offensive language. Then too it is partially confected, in that the real diary was condensed and reshaped into a series of letters for the purpose of publication. It cannot in the remotest way be equated to the testimony of a person who experienced slavery, such as Solomon Northrup's *Twelve Years a Slave*. What it does show, however, is what a white plantation owner knew about the system from which they extracted such idle, pleasure-filled lives. Rape, torture, death: none of the horrors that Fanny reported to her husband came as a surprise, and nor did they challenge his commitment to slavery, though he understood that not everyone would view it as he did. When she left him in horror and disgust, he took their two daughters and refused to grant her access as a way of coercing her into keeping silent, knowing that publication of the diaries could drive the cause of abolition in America. Again, the punishment Fanny underwent is minuscule in comparison to what the women who could not leave Butler Island experienced, though part of how we know about their particular losses in such detail – the rapes, miscarriages, still-births and infant deaths – is because of Fanny Kemble's diary.

This inflammatory document remained unpublished until 1863, the midpoint of the American Civil War. Fanny's daughters were both over twenty-one by then, old enough to see her without their father's permission, and she hoped that an unvarnished account of slavery might help unsettle Britain's support for the Confederacy. It was published in England in late May, while Parliament was debating the question of recognising the Confederacy, and in America on 16 July, priced at a dollar and a quarter. It's often said now that its publication changed the course of the war, or that it successfully undermined the South in British minds. In fact, what British reviews Kemble received were mixed. While the American *Atlantic* praised 'the first ample, lucid, faithful, detailed account, from the actual head-quarters of a slave-plantation in this country, of the workings of the system', the English *Saturday Review* was appalled, not by the terrible things that Kemble had seen, but by the fact that she, a woman, thought it appropriate to share them.

> The particulars into which she enters betray an ignorance of English notions of refinement quite marvellous and which will make her readers absolutely stare. The coolness with which she prints sundry details which few ladies would put on paper even in manuscript unless called upon to do so by the necessities of the case is quite inimitable . . . We never met with such minutiae in print except in the pages of professedly medical publications, at least in books of the present day; and however well meant may be their insertion in a lady's journal, they will cause many a sudden 'pull up' in drawing rooms where the book may be read aloud for the amusement of a family group.

What this statement makes clear is that Kemble had broken the taboo of silence – silence about women's bodies, about violence, rape, menstruation, prolapsed uteruses, infections. Above all, silence about the origins of money. It also testifies to a Britain, post-abolition, where a family might regard an account of a slave plantation as suitable drawing-room entertainment, though perhaps ideally with some titillating redaction of the more disturbing words.

By the time *Journal of a Residence* was published, William Fowle Fowle Middleton was dead, and the Middleton wealth in America was melting away. Emancipation was coming, whether the planters wanted it or not. Two scenes in particular encapsulate those years of reckoning. The first is on the night of 2 June 1863, when the former slave and antislavery activist Harriet Tubman led Union soldiers on a raid down the Combahee River. Tubman had spent years guiding slaves out of the south into freedom by way of the Underground Railroad, and now she navigated three tugboats past the Confederate mines. That night, the soldiers burned down many plantations, liberating perhaps seven hundred slaves. One of these plantations, Newport, belonged to members of the Middleton family. The overseer's house was destroyed and one hundred and thirty people escaped into freedom, though the Middletons remained convinced that they'd been driven off by Union soldiers. They didn't see what Tubman did: the people surging joyfully onto the boats, carrying cooking pots and pigs. She never forgot the sound they made. 'We laughed, an' laughed, an' laughed.'

On 23 February 1865, Middleton Place too was burned down by Union soldiers. A family member reported that the newly freed slaves ate dinner in the dining room and then went to the mausoleum and scattered the Middletons' bones across that exquisite green lawn. This might be racist propaganda, in keeping with the virulent and hateful way the Middletons talked about the people they'd once owned over the decades of Reconstruction and beyond. On the other hand, perhaps it was the right epitaph for a garden that had been founded on so much cruelty and pain, to state without artifice that it was a boneyard, where the ties of blood and family had been severed for so long.

I would be glad to end with that laughter. There are gardens that have come at far too high a price, that should never have been made, and I am glad that Crowfield is now obliterated, and that the historians at Middleton Place have tried to recover and foreground the stories of the enslaved people who built and paid for its garden, with its rare camellias and azaleas. Beauty is not a virtue that floats free of cost. Not to my mind, anyway. Whatever Shrubland was, it wasn't beautiful.

When I was looking at the family letters in the Suffolk archive, where they'd been deposited in lieu of death duty, I came across a cache of old photos of the garden, taken in 1908. It was all there, just as Adveno Brooke had described: the classical busts, the yew hedges, the monumental temples and pavilions, decorated with nymphs and sets of antlers, symbolising conquest both classical and natural. In nearly all of these photos there were no people, and the emptiness contrasted strangely with the excess of ornamentation. Who was this exquisite landscape for? How could it possibly have been considered worth the cost in human suffering? The Swiss Cottage was deserted. The pool reflected empty sky. I kept going through them, hypnotised by the spectacle. It was as if an entire dynasty had upped and left, abandoning their fountains, their cypress trees. Shrubland had become its own memorial, a shrine to futility and greed. There are better ways to make a garden.

V

GARDEN STATE

Birds from the moment we woke up. The light kept expanding, the afternoons opening at the latch, a dreamy feeling at half past three, then four. Spring came in a wash of colour, each day's tide subtly different from the last. I was out at first light in pyjamas and a coat, poking about the beds, looking for new arrivals. The hellebores came first, glowing white cups flushed pink or green at their margins, then pink stippled with mauve dots, or strangely suffused with green, their organs protruding. There were deep wine-red ones pointed like stars, a freckled burgundy, a satiny near-black, and my favourite, a porcelain richly veined with maroon, with lovely rounded petals. I took them into the house and floated them in the cut-glass bowl my grandmother used for trifle, and in the heat they opened up and fell into new patterns, like indolent swimmers.

There were more and more snowdrops too, piling up beneath the trees. I found a clump of doubles beside the covered well. It was 'Flore Pleno', favourite of the Victorians, its layers of porcelain skirts finely marked with hair-thin green lines. By

5 February it was so warm I drank my coffee in the garden, listening as the bees swarmed over the daphne, wobbly and roaring after a surfeit of sugar. The house was intersected by new streams and currents of light. Then the temperature dropped from 10 degrees to -6 overnight and by morning a foot of snow had obliterated the garden.

I love the sense of snow falling in the dark. As the demarcation between beds and lawns was erased, the garden's structure came to the fore. It was reduced to a sinuous exoskeleton, a complex armature of curves and archways, its buttresses and piers thickly capped in white. The only colour now came from the witch hazel, which flowered on unconcerned, its dangling petals like zested strips of lemon rind, producing a smell so hypnotic I kept snapping off twigs to scent the house. Walks were perilous. Ian fell hip-deep into a ditch of snow and drove home soaking. After a tethered week the thaw was a relief. I went out at sunset to see how the garden had stood up. *4:55*, my diary records. Everything felt tremulous. There were tiny flies whirling above the yucca, and somewhere out of sight a blackbird singing beneath a fingernail of moon.

After this false start, my lists of the day's new flowers began to swell. There weren't as many crocuses as I'd hoped, but dozens of primroses in dull pinks and milky yellows, and then a wonderful deep carmine red. In the neglected bed under the yew hedge I found a blue pulmonaria, maybe 'Blue Ensign', surrounded by *Arum italicum*, the marbling on its leaves like snail trails. Were the alien, shiny, thrusting spikes that had appeared in the rose border crown imperials? They were, flowering fox-orange and reeking at the end of the month. The ribes beat them to it, each fat bud unspringing a glossy white raceme. The smell of warm earth was so intoxicating I wanted to roll on the grass in my jacket.

On one of my morning patrols I saw a tiny scrap of blue in the pond beds, darker this time than the little iris I'd found before Christmas. It was the scilla Mark had written about, the Prussian blue *Scilla siberica* 'Spring Beauty', followed by the paler *Scilla luciliae*, which used to be called *Chionodoxa*, identified by a touch of white at its throat. Not very many, I complained, but within a fortnight it was as if the empty beds had filled with a thin trickle of water that caught and gave back the sky.

It sounds as if all I was doing was looking, and it's true that sometimes I took a flower into the house so that I could gaze at it beneath Ian's magnifying glass, the better to assess its adorning frills or stripes or splashes. But hours each day were still spent clearing and mulching, transplanting foxgloves, potting up dahlias. As for the long evenings, I spent them hunched over my computer, picking through census records. Back in the previous spring, after we'd seen the house but before we'd moved in, I'd spent a lot of time hunting online for references to Mark. While I was voyaging round Google I came across the house's name in *The Women's Suffrage Movement: A Reference Guide, 1866–1928*. I'd clicked the link, of course, and so had encountered for the first time the utopian socialists Goodwyn and Catherine Barmby, builders of a Victorian Eden that couldn't have been further from the carpet bedding and cruelty of Shrublands Hall.

In the clutter of the move I'd forgotten about the Barmbys, but spring brought them back to mind. Goodwyn Barmby was born in our village in 1820. He was meant to enter the Church of England, but his father's death when he was fourteen freed him from this obligation. He doesn't seem to have gone to school for long, instead reading broadly under his own direction. His father had been a solicitor, and the family were of modest means and intense intellectual pursuits. Years later, his mother's will included boxes of fossils and books on history, geology, conchology and mineralogy; a very Victorian set of interests.

At sixteen Barmby was lecturing local agricultural labourers in his soft, persuasive voice about the new Poor Law Amendment Act, which saw the impoverished confined in workhouses, where conditions were deliberately made so appalling that only the most destitute would ask for help. The Act represented another tightening of the screws upon a rural class already devastated by enclosure. At seventeen he moved to London and lived among revolutionaries, and at twenty he went to Paris to meet with utopian socialists, from which trip he returned with a new word for the English language: *communism*.

Barmby came of age amidst a tumult of new thoughts and ideas. He was at different and sometimes overlapping periods a Chartist, a utopian socialist, a feminist, a millenarian, a Christian, a visionary – as well, it must be admitted, as a pretentious crank. He wanted to re-establish Eden, and indeed believed the original Adam and Eve were one hermaphrodite being, who had been most unhappily split asunder, and that in the future humans would relinquish the false polarity of gender and embrace the masculine and feminine elements at work in every being. He seems to have had almost limitless energy, founding several journals, for which he tended to write all the articles, producing reams of cod-Shelleyean poetry, preaching, lecturing, and even establishing his own church. As his one-time co-editor Thomas Frost recalled, 'he hurried on, with a zigzag course, in a manner which made it ever doubtful whether he would be found at the point where he had been heard of last.'

In 1841 he was selected as the Chartist parliamentary candidate for Ipswich, the same constituency that William Middleton had once represented. The Chartist struggle, then reaching its zenith, was concerned with winning the vote for working-class men, along with anti-corruption reforms of parliament. Barmby represented the East Suffolk Chartist Council at the National Convention in 1839 and 1841 but his ideas were already larger and more far-reaching. He wanted women's suffrage too, and marriage reform, as well as the wholesale liberation of women as the economic and emotional supplicants of men. But he also wanted to change the order on which the world was founded, not just correct its malfunctioning parts, rebuilding it from the ground up as a place of radical equality and communal love.

In this he was influenced by Robert Owen, the Welsh factory owner who made a fortune from his cotton mills but came to see capitalism as a curse that drove individualism, bringing mass suffering as the price for wealth. Owen sought abolition of the whole damned regime: marriage, religion, private property and all, in favour of radical democracy, equality and communitarianism. He thought how people behaved – their nature, if you like – was not fixed, as the Church suggested, but shaped and constrained by their circumstances. If capitalism produced competition and with it inevitable failure and subjugation, as well as the isolation brought about by avarice, then new circumstances were required, in which private accumulation was unknown. To this end, Owen proposed the establishment of a radical new model of organisation: self-sufficient, land-based communities where both work and child-care were shared endeavours.

In 1841 Barmby married a fellow Owenite, Catherine Watkins, who had been publishing passionate defences of women's rights in the Owenite newspaper *The New Moral World*, under the pen-name Kate. Together they established what seems to have been the first campaign for women's suffrage, a goal the Chartists were not willing to address. The Barmbys' Declaration of Electoral Reform asserted the equality of men and women. 'We are opposed to sex legislation as we are opposed to class legislation. We therefore ask by the names of Mary Wolstonecraft' – whose name they spelled wrong – 'and Charlotte Corday, for universal suffrage to woman and man, for unsexual Chartism!'

That year, which Barmby grandiosely declared Year 1 of the new Communist Calendar, they also set up the Communist Church, an institution that fused their religious and political aspirations. In its first, pre-Marxist sense, the word 'communism' meant communal or communitarian, and did not have the same totalising vision we associate with it today. The Barmby

Communists were, as Barbara Taylor puts it in *Eve and the New Jerusalem*, 'a highly cerebral and intensely romantic lot, vigorously experimental in their approach to daily life.' They engaged in meditation, hypnotherapy and vegetarianism, just like any self-respecting Sixties commune. Barmby grew his golden hair long and parted in the centre, and sported a Byronic necktie. The couple travelled London in a hooded cart, handing out tracts that opposed the payment of taxes and demanded the resignation of the Archbishop of Canterbury. Sometimes they invaded churches, lecturing the congregation on the evils of trade.

Barmby had long fantasised about the ideal community that might arise following a widespread adoption of Communist ideals, furnishing it with libraries, fountains, airships, as well as buildings of 'the most gorgeous conceptions and glorious imaginings, which even the most advanced among us cannot at present conceive or imagine.' He knew it wouldn't materialise overnight, and so in 1842 began to prepare the ground in Hanwell, Middlesex, a place best known for its lunatic asylum. There were no servants at the Morville Communitorium, and what acolytes could be found soon melted away when confronted with the rigorous demands and privations of communal life. If utopia depended on shared housework, not everyone was willing to pay.

The Communitorium lasted a single year. Nothing burns faster than youthful ideals, and after Catherine died of consumption on Boxing Day 1853, Barmby became a more conventional Unitarian minister, though he never stopped agitating for women's suffrage. He spent most of his middle age in Wakefield, accompanied by a new wife, Ada, and his suffragist sister Julia. Both women were among the 1,499 signatories of the 1866 Women's Suffrage petition. When Barmby's mother died in 1879, leaving her children shares in two collieries in Yorkshire, as well as books on flowers, her writing case and oil paints, a snuff box and £24 in cash, this radical family returned to their childhood home in Suffolk, where Barmby died soon after.

Dredging through the 1891 census, I found Julia and two more sisters, Charlotte and Clara, had subsequently moved to our house, a minute or two's walk from where they'd grown up. I liked to think of them there, reading the poems in the commonplace books that had somehow ended up in a storage room at Yale, or discussing the right to vote over little lustre-ware cups of Darjeeling tea. They must certainly have walked in the garden, though the only tree I was sure they'd have seen was the mulberry, which must even then have been two centuries old. Their presence tied the garden lightly into the long history of utopia, and it also reminded me of my own utopian experiments, which had been established for some of the same reasons as the Morville Communitorium and had foundered on some of the same pervasive rocks.

In the 1990s I'd been involved in environmental protest and then decided that resistance was no good, or not good enough; that the important thing was to establish new ways of living. The struggle: the struggle is always filled with these consider-ations, about the best position from which to effect change. I've never been a person especially comfortable in a group. I prefer to be outside, following my own nose, listening hard and thinking privately, but in those years I was always in co-ops and covens and affinity groups. One hundred and fifty years after the Barmbys, and the tools had barely changed. Newsletters, public meetings, communities, new projects that burned with excitement and then guttered out, since some people were restless advocates who did no work and others – but I'm sure you've experienced group dynamics too.

In the spring of 1999, I moved into a women's housing co-op on the outskirts of Brighton. In those fin de siècle days before the internet infiltrated and reshaped our lives, Brighton was an alternative outpost akin to San Francisco, where all manner of hippy activities remained in fitful animation, decades after the Summer of Love had been rained off elsewhere. The house was on the Moulsecoomb council estate but had the primitive atmosphere of Cold Comfort Farm: a dense fug composed of Nag Champa incense, wood smoke and washable sanitary towels. A census-taker assessing regional poverty levels was shocked to learn that we didn't have a washing machine, tumble dryer, toaster, or television. It wasn't that kind of scene. We made vegan ice cream to sell at raves and bonded at full moon with bare-breasted rituals in the back garden, much to the amusement of our neighbours.

The co-op was called Hecate, after the Greek goddess of witchcraft. At a rough estimate half its members had arrived by way of feminist or women's groups in the city, and the rest, me included, from environmental activism, drifting in from traveller sites or road protests. I arrived muddy-booted from a spell living outdoors during a long, isolated winter on another utopian concern: the Natty Trust, whose alluring-sounding mission was to take over degraded land and rehabilitate it sustainably for public benefit. The first of these sites was an abandoned pig farm on the outskirts of Brighton, tucked under the lee of Wolstonbury Hill. An assortment of ramshackle barns and sheds led into a ten-acre meadow. There was an idea that many people would live there, but that winter it was just me in a bender, a kind of rudimentary tent I made myself of bent hazel poles and a tarpaulin, cooking on a fire and sleeping with an axe under my pillow.

The appeal of both Hecate and the Natty Trust was that they were positive ventures. I was tired of living somewhere solely to prevent it from being destroyed. I wanted permanence: an experimental ecologically-minded community that could provide the best aspects of protest life – low-impact and close to nature – without the constant threat of destruction and eviction. This desire was widely shared among activists I knew. Friends ended up at Tinker's Bubble in Somerset, Tipi Valley in Llandeilo, the Centre for Alternative Technologies in Machynlleth; even at Findhorn. All bar Tinker's Bubble dated from the 1970s, stubborn remnants of a wave of utopian dreaming that had otherwise been obliterated by Thatcherism.

Highly cerebral and highly romantic: Barbara Taylor's description of the Barmbys fitted me too. My twenties were marked by an inability to finish anything, to put a superfluity of half-formed ideas into practice, with the exception of studying and then practising herbal medicine, work that consumed the entire decade. Plants were always spilling from the centre of any utopia I imagined. I loved the idea of a communal or remedial garden. My notebooks were illuminated with pictures torn from *National Geographic*, of gardens in hospitals and prisons, or meadows on the Appalachian trail. The plants themselves encoded something utopian for me, a vision of how the world ought to be. At the same time I had a disinclination to act so powerful it felt ordained. It wasn't just laziness; it was a sense of the pointlessness of human effort, that almost every action had some damaging consequence, and that it was better to simply let things be.

I did a permaculture course at Plants for a Future, but I don't remember once digging with a spade at Priestfield, though I spent hours identifying the plants that grew there, thriving on neglect. Fescue and agrimony, ribwort plantain and the little ground ivy, *Glechoma hederacea*, which is related to mint and serves as a spring tonic. I started gardening at the Hecate house, but this too was ineffectual. I was beaten back by the rubbish dump we discovered just beneath the thin soil. It wasn't a garden at all, really, just a fenced segment of the chalky Downs, and I didn't have the knowledge or energy then to transform it into the kind of paradise I longed for, the cottage or monastery gardens that I found so persistently desirable.

Staying power had come later, in my thirties; the helpful revelation that there is no mystery to any creative endeavour, save for turning up each day and knuckling down to it, hour after tedious or – but this was far rarer – joyful hour. That was when I started making gardens and writing books, the unhappy years of failed dreaming an increasingly unvisited memory. The Barmbys had brought it back, my stint in the Morville Communitorium, a place I'm glad I visited and gladder still I left.

Three bare-root roses soaked in a bucket while I crossed back and forth, digging holes and sprinkling libations of chicken manure pellets and mycorrhizal fungi. I put 'Shropshire Lass' on the north wall, crossing my fingers about the lack of light, and the striped pink and carmine *Rosa mundi* in the yew border. It was Derek Jarman's favourite rose, a sport of *Rosa gallica* that dates back to the twelfth century, the crumpled flowers reminiscent of carved bosses from a medieval ceiling or a William Morris blockprint. Then there was 'Mermaid' to fill a gap on the high curved wall at the end of the pond garden, between the robinia and a fig. The wall there was made of very soft red brick, pierced with hundreds of holes by mining bees, and I liked the idea of them emerging through a lattice of loose canary-yellow flowers, each with a heavy fringed eye of golden stamens, like a cow's lashes.

Paradise haunts gardens, Jarman once said, but it is also true to say that gardens haunt utopias. The two are close associates, no doubt in part because the inextricable connection between gardens and paradise places them so far up the hierarchy of desirable attributes and conditions for an ideal society, as well as one of its abiding metaphors. In More's *Utopia*, for example, the citizens of Amaurot have finely tended gardens, models of beauty and utility, full of herbs, vines, fruits and flowers. There is no conception of private property, so that anyone can enter any home at will, while every ten years all the residents are assigned a new house by lot. Despite this transience, they compete to make the best garden, and this competition is the reason, More says, that Amaurot is such a beautiful city.

In William Morris's *News from Nowhere*, meanwhile, gardens running with flowers are one of the many delights encountered by the time-travelling narrator William Guest, who goes to sleep in Victorian England and wakes in a twenty-first century totally reconfigured after a socialist revolution. Throughout his journey, Guest is astounded to find himself in familiar regions of London that have undergone a deliciously floral transformation, so that Kensington is a forest, Trafalgar Square an orchard of apricot trees and Endell Street is filled with roses. In fact, the dominant image of this new society, if the word dominant can be used for such a gentle civilisation, is that of a garden, 'where nothing is wasted and nothing is spoilt.'

What makes a garden such an important constituent of a utopia? It is neither a farm nor a wilderness, though it can push up hard against either of these extremes. This means it betokens more than just utility, encompassing beauty, pleasure and delight, while remaining emphatically a site of labour as well as leisure, a place to please puritans and sybarites alike. The presence of gardens in a society indicates that its inhabitants have sufficient surplus energy and time to attend to cultivation that, like art-making, is not strictly speaking necessary. What's more, they wish to do so, which perhaps conveys something positive about their emotional or even spiritual state (which isn't to say there aren't gardens constructed in rage or sorrow). A garden is expressive of private fancy, while generating a superfluity of beauty. If a new model of society is desired, one that attempts to share its burdens and benefits more equably, then the question of the garden becomes very interesting to contemplate.

I'm not sure any utopian dreamer has valued a garden more than Morris, that burly Victorian visionary who worked with such inexhaustible vigour at making a society that was both just and lovely. Morris was so sure of beauty as a virtue, not a luxury, even though his own political work was sometimes underpinned by exactly the kind of capitalist enterprise he despised. We inhabit a society right now that takes pleasure in rejecting this kind of complexity out of hand, but I think the ambiguities of Morris's position make him even more useful as a guide to how the garden of utopia might be planted, since we need to start from our contaminated present and not some future position of undiluted purity.

When I was a child in the 1970s, houses were still full of Morris & Co. prints: nostalgic indoor gardens, their walls, curtains and armchairs all swimming with willow boughs and pomegranates, rabbits and strawberries, at once impeccably polite and oddly disturbing. My earliest memory of a pattern is a sofa upholstered with a Sanderson Morris print that dates back to when my father was still living with us, before my sister was even born. It was a version of the famous Golden Lily design, one of the many Morris and Co. designs made by

J. H. Dearle. The stylised flowers were coloured in shades of nutmeg, tobacco and raw sienna, with stippled petals and thrusting stamens striped like Twizzlers. As a child I found it both oppressive and alluring, somehow simultaneously comforting and claustrophobic. There was no escape from design. Even the ochre-coloured ground was filled with darker curlicues, fluent shadows of forms that while floral in origin were by no means botanically accurate or even always identifiable, making them strangely less restful, as if the mutations and proliferations might never stop.

By the 1980s Morris was far out of fashion and the old sofa was reupholstered. It eventually moved with me to Brighton, where I lived for a time around the corner from a beautiful, etiolated girl with something of the looks and comportment of a medieval page. She'd grown up in Kelmscott, Morris's house by the Thames in Hammersmith, itself the model for the house in which William Guest slips into the future, waking in a new century in *News from Nowhere*. I don't suppose I was very interested at the time. Nothing Victorian felt alluring then, though it's clear to me now that the anti-roads movement I was involved in had roots in the anti-industrialism and proto-ecology of Morris and Ruskin. We were more taken by the Diggers, the breakaway sect of the English Civil War, who occupied common land, preaching a more radical agenda of wealth redistribution than Cromwell and his men. To turn this story full circle, the Diggers were driven from several commons by the landlords they wished to depose, the last of which was in Iver in Buckinghamshire, where I lived as a child with my parents, and where I was photographed on the Morris sofa, holding my new-born sister amidst a golden garden of strange blooms.

It pleases me, this plumb-line dropped through time, since Morris and the Diggers were engaged in a similar sort of revolutionary work, which we might broadly call reclaiming the commons. The Diggers, also known as the True Levellers, were ordinary people, often on the brink of starvation. Like Milton, they hoped that the execution of the king in January 1649 was the first step in the creation of a more equable new order. This was the great era of tract- and pamphlet-writing, and the Diggers produced them in their dozens, drawing on scripture and especially the Book of Genesis in defence of their case. Many were written by their leader, Gerrard Winstanley, best remembered now for declaring the earth to be a 'common treasury', given by God equally to all men and never intended to be bought or sold; an early exemplar of what Barmby would later call communism.

Winstanley claimed a vision directed him to establish the first Digger community on St George's Hill on 1 April 1649, planting it with carrots and corn, while a voice heard 'in trance and out of trance' told him that the common land should be tended in common, its fruits shared by all who laboured on it. This same part-mystical, part-lawyerly argument is pursued in *A Light Shining in Buckinghamshire* and the rather less snappily titled *A Declaration of the Grounds and Reasons, why we the poor Inhabitants of the parish of Iver in Buckinghamshire, have begun to dig and manure the common and waste Land belonging to the aforesaid Inhabitants, and there are many more that give consent,* which was published the following spring, after several Digger communities had already been destroyed, their crops torn up, their crude dwellings smashed and their residents beaten and arrested.

There is, writes the anonymous author of *A Declaration of the Grounds and Reasons*, no 'righteous power' to sell or give the earth away, making the landlords with their enclosures not only iniquitous and cruel, but acting against the will of God. Sample this singular sentence, which winds itself pleadingly, forcibly around the issue:

> We are urged to go forth and Act in this righteous work, because of our present necessity, and want of the comfort which belongs to our Creation, that the earth being inclosed into the hands of a few, whereby time, custome and usurping Laws have made particular Interests for some, and not for all: so that these great Taskmasters will allow us none of the earth whilst we are alive, but onely when we are dead, they will afford us just as much as will make the length of our graves, because they cannot then keep it from us, and that then we should be equall with them; but why may we not whilst we are alive with them, have as much of the earth as themselves?

The Diggers lost, of course, and the common land and wastes were enclosed not for the benefit of the poor, as Winstanley wanted, but for the rich. John Clare, who would have appreciated that bitter line about only the dead being allocated an equality of land, could have told them how much worse things would get.

Two hundred years after the English revolution, and without reading either Winstanley or Clare, both of whose work spent many years burning in the dark, unpublished and unread, William Morris came to believe something similar about the land and how it should be shared, though he could hardly have emerged in more different conditions. He was born in Walthamstow in 1834, and raised for the most part in a Palladian mansion on the edge of Epping Forest, with a fifty-acre garden and another hundred acres of farmland in which to walk and fish and ride on his Shetland pony: a little princeling in a suit of model armour, lonely, curious and very confident of his status.

Melon beds, white-currant bushes, peaches ripening on a sun-warmed wall, each with their own distinctive odour, which would carry and encode memories for decades to come. He made a child's garden within this immense and productive enclave, and in the dim evenings pored over Gerard's *Herball*, its bewitching woodcuts made in Antwerp by a succession of Renaissance artisans, who cut the forms of hellebores and wild thyme into blocks of pear wood, which were shipped to London for printing. If the garden meant plenitude, abundance, over-brimming sensual delight, then these printed flowers offered a way to store and transport it. The origin of Morris's adult vision lies surely in these two abundant gardens, one paper and one real.

Like Barmby, who in many ways represents his more hopeless and quixotic twin, Morris's father died when he was still a boy. William Morris Sr. was a solicitor who speculated lucratively in Devon Great Consols, investing at the very start of what soon became one of the most productive copper mines in the world. As the quality of the copper declined, the mine transitioned into arsenic production, providing by 1870 half the world's supply. These canny, ecologically deleterious shares kept Mrs Morris and her children afloat, if not so rich as to be able to stay on at Woodford Hall. Later, a directorship at the mine provided the adult William with an income that gave him freedom to engage in creative projects of many kinds. He finally divested himself of an active role in 1875, sitting firmly on his top hat to underscore his rejection of the bourgeoisie, not that he'd ever much liked being togged out in formal clothes.

It took me a long time to come round to Morris. He was too closely associated in my mind with drippy pre-Raphaelite paintings of damsels in girdles and chivalrous, grieving knights. Somehow, he was too large to grasp, so gifted in so many arenas that he seemed almost glib, suspiciously fluent and perhaps without depth. I was wrong about that. It's true that he was a dynamo, a spinning top, who compulsively taught himself to master a dozen crafts, who could weave a tapestry and dye a chintz, embroider a wall-hanging, construct a stained-glass window, write a poem (often on the bus and often too at the astounding rate of a thousand lines a day), illuminate a manuscript, bind and print a book, perhaps translating Homer or Virgil as an evening's respite from this more exacting work.

Morris with his wild curls and sea captain's beard, plump, shy, open-handed, electric, chanting his own poetry for hours

at a time as his friends stifled their yawns and drew cruel cari-
catures of buttons bursting over his full belly, perhaps irritated
by his unquenchable earnestness and energy. Morris laughing
as he worked, periodically going off into stamping rages like
Rumpelstiltskin; Morris the cuckold, whose wife Janey carried
out a long love affair with one of his closest friends; Morris the
revolutionary, marching with striking miners, dressed in his
blue serge jacket, delivering the new-pressed message of
socialism to unheated meeting halls and street corners all round
the country: a prodigious man, in short, both in his capacity for
work and for imagining out of the lack he saw and felt a better,
more beautiful world, with no division between rich and poor.

Gardens were part of his conception of that world, as well
as a metaphor for conveying its many riches. They were at the
heart of his design work, but more importantly they formed the
centre of his vision for how society itself could be transformed.
He only created a handful himself, but a surprising amount
emerged from out of them, their tendrils extending luxuriantly
right into the present day. The first was at Red House, the home
designed for him by his architect friend Philip Webb. It was
built in an acre of Kentish orchard in what was already starting
to become the London suburb of Bexleyheath, so closely set
amid the old trees that on warm autumn days Russets and
Pearmains would bounce in through the windows. Many things
started for Morris in this idiosyncratic, secretive house, its
interior fantastically decorated with wall paintings and embroidered hangings. It's here that he began his married life with
Janey, that their two daughters May and Jenny were born, and

that he founded a design co-operative with a group of artist friends, among them Dante Gabriel Rossetti and Edward Burne-Jones, the famous Firm that would later become Morris and Co.

From the beginning Red House was designed to be infiltrated by the outside. Webb's plans even specified the climbers for the west wall, fragrant arcs of jasmine, honeysuckle, passionflower and roses. The house was L-shaped, so that it both contained and was enfolded by the garden, which Morris planted with cottage garden, almost childish flowers: lavender and rosemary, poppies, sunflowers and lilies. He laid it out as a series of small secluded rooms called *plaisances*, with arbours and wattle fences covered in roses to mark out the divisions. It was closely modelled on the medieval hortus conclusus, the enclosed monastery or castle gardens that he'd pored over in fifteenth-century manuscripts and books of hours at the Bodleian Library and the British Museum, often with the Virgin Mary at its centre.

The garden at Red House represented a distinct shift in style, without any precedent in Morris's own time. It was a wholesale rejection of the landscape garden and its classical antecedents, as well as the power relations it encoded. If Capability Brown extended the garden into the landscape, expensively reshaping natural features to produce pictorial views and dominating vistas, what Morris offered was an explicitly medieval vision of the garden as small-scale, intimate, cultivated and domestic.

A garden of any size, he explained in his lecture 'Making the

Best of It', 'should look both orderly and rich. It should be well fenced from the outside world. It should by no means imitate either the wilfulness or wildness of Nature, but should look like a thing never to be seen except near a house. It should, in fact, look like part of the house.' As for plants, here too he parted from the prevailing Victorian fashion, favouring old-fashioned simples and singles over the artificially coloured, the grotesque doubles and over-bred mop-heads. He spoke with a particular disdain of carpet bedding, the massed scarlet geraniums and yellow calceolarias of a Shrubland Hall firmly dismissed as 'an aberration of the human mind.' His ideas were absorbed by the coming generation of Arts and Crafts gardeners, from William Robinson, who likewise loathed carpet bedding, to Gertrude Jekyll, who as a young garden designer made a pilgrimage to visit Morris. The garden as a succession of living rooms in which to live is almost the defining feature of some of the twentieth century's most beautiful Arts and Crafts gardens, from Hidcote to Sissinghurst to Great Dixter, but it was Morris the medievalist who first planted out the idea.

As well as revolutionising real gardens, he was forever inviting the garden inside, obsessively filling his buildings and furniture and fabrics and books with a non-stop proliferation of plant life, an unceasing, shape-shifting, tidal wave of flowers, by turns innocent, sensual, erotic, and downright disturbing. In this way, the wild was made domestic and the domestic wild, and Morris's gardens became a fairy-tale kingdom, a world apart. By the 1870s, he was even using plants to manufacture as well as feature in his designs, refusing harsh chemical aniline dyes and instead reverting painstakingly to what had gone before, drawing on Gerard as his source. Brown from walnut husks, red from madder root, blue from indigo, yellow from weld and marigold and poplar twigs: those flowing chintzes were floral twice over, though the madder periodically polluted the river at Morris's factory, a tributary of his beloved Thames.

Part of the pervasive magic of Red House was its isolation, but Morris's commute to London soon became exhausting. Janey was often ill and by 1865 he'd decided to relinquish the house in favour of a more practical London residence that could double as a business premises. That autumn the family moved to 26 Queen Square, a four-storey townhouse on a site now occupied, oddly enough, by the wing of the National Hospital for Neurology and Neurosurgery where my step-mother died. It was in Queen Square garden that my father had sat, holding her belongings in a cardboard box as he wept down the phone to me the week we moved, a memory that felt increasingly bitter as the struggle over his house dragged on and on.

This public square, planted with plane trees, was the only garden available to the inhabitants of No. 26 and by the mid-1870s the Morris family had moved again, dividing their time in various configurations between two rented houses, one rural and one urban. At Kelmscott House in Hammersmith there was a garden with a greenhouse ('real', Morris announced in a persuasive letter of discovery) and another orchard, with apple and pear blossom in the spring and dishes of mulberries come August. The hyacinths were very fine, though the daffodils disappointed him by coming up blind. Like all London gardens it had a tendency to feel dreary and sooty when the weather was bad, but Morris's letters to his daughter Jenny, his chief correspondent in those years, persevere at winkling out its charms, among them Japanese anemones and what he teasingly calls 'Chaynee oysters', better known as asters.

The real delight of Kelmscott House was that it was on the Thames, the front looking over the water and the garden set behind. The river connected Morris with his other and far more beloved home, Kelmscott Manor in Oxfordshire, which he called Heaven on Earth. Both houses appear in *News from Nowhere*, like bookends. Guest wakes in his house in Hammersmith, and at the conclusion of his adventures the company travel up the Thames by boat to a version of Kelmscott Manor. There Guest enters a garden so lush that, in one of Morris's most memorable images, the roses 'were rolling over one another.' An illustration of the manor appears on the frontispiece too, pigeons clapping overhead, those super-abundant standard roses lining the flagstone path.

If Red House was inspired by Morris's fantasies of the past, Kelmscott was the real thing, a multi-gabled grey Elizabethan manor house, built in the late sixteenth century and arising from a seemingly untouched pastoral landscape. The garden was straight out of Gerard, with its violets and winter aconites, its meads of tulips and fritillary, its tumbling strawberry beds infiltrated by hollyhocks and raided by thrushes, despite the nets. Many of Morris's most famous designs originated at the manor, carrying its teeming atmosphere to nurseries and sitting rooms all across the world. Something sorrowful or yearning, too. Kelmscott was shared, after all, with Morris's difficult friend Dante Gabriel Rossetti, the lover of his wife.

My own garden was quickening by the hour. I mowed the grass for the first time on 29 March, cutting a curved path through the greenhouse lawn. A bat in the blue air, I wrote in my diary that evening, and later a field mouse on the doorstep. The next morning there were jackdaws in the pigeon loft, their beaks stuffed with twigs, shouting out their quarrelsome *chack chack chack*. Spring had swung into gear. We ate the first breakfast outdoors, and afterwards I filled the pots in the pond garden with pelargoniums: 'Surcouf', its flowers the startling pink of pickled beetroot, and 'Crocodile', named for its scaly variegated leaves.

There was a triangular abandoned bed on the north side of the house, across the path from the larger, sunnier one Ian had claimed as his herb garden. A cheerful winter jasmine grew out of gravel, under which there was one of those horrible weed membranes made of black plastic, which inevitably degrades into ribbons and will outlive us all. Morris would have hated it. On an impulse that afternoon I pulled it out, planting the gravel instead with transplanted violets and primroses and a foetid hellebore I found languishing beside the bins.

It was the hottest March day in thirty-five years. I ate a hot cross bun and drank a cup of tea, and then went at the bed under the sitting-room window, which had also been covered in gravel and planted with low-maintenance lavender, now straggly and sad. By the end of the day, I had a new bed turned over and ready for planting. *The 'Vanessa Bell' rose?*, I wrote. *Pinks, pulsatilla, Shirley poppies?* That was the thing about gardening. The possibilities were infinite. A few hours later, I was back to scribble: *Bats again.*

I was beginning to feel drunk on the sheer variety of surviving forms. White violets, pink bergenia, the sappy, cadmium-green spikes of day lilies, and then the true harbinger of spring: tissue-pink magnolia petals emerging from their purses, leaving velvety husks on the grass. At first it was only two or three, and the next morning it was as if a great ship had unfurled its sails and was riding at anchor on the lawn. The phrase *so much to do* appeared on nearly every page of my diary, along with *fiddling about*, which often consumed whole days. I was pruning the beech hedge when they rang the passing bell for Prince Philip, and painting the greenhouse after my first Covid vaccine. Some mornings I was up and out by four or four-thirty: tugging up cleavers; building tripods and planting out sweet peas; clipping the box squares, still in my pyjamas, dipping the shears into dilute bleach to ward off blight.

By my birthday in the middle of April it was as if I'd climbed inside a Morris pattern. The shabby little greenhouse garden had transformed itself into the meadow I'd planned and planted back in winter. It was at its most beautiful early in the mornings and just before sunset, when the low sun filtered through the enclosing green walls of the hornbeam hedge, illuminating the chequered fritillaries, with their foxed purple skins, so they looked like the shining forms in a Turkish carpet. In the pond

beds a rising tide of bluebells and Welsh poppies was interrupted by upsurging fountains of glaucous cardoon leaves. A
little mossy standard-trained *Viburnum carlesii* I'd ignored all
winter produced absurd pink pompoms, exuding intoxicating
rivulets of scent. Matt came and hauled the rambling rose out
of the cherry, which was finally due for the chop. I left him with
a mug of Earl Grey and came back a few hours later to find
he'd managed to cajole the stiff old stems into loops and curlicues that covered the whole back wall.

As bare patches were revealed I started to draw up embryonic planting plans. I wanted a group of *Narcissus cyclamineus*
under the magnolia, the ones that look like Piglet with his ears
blown back. Acid yellow, ballet pink. More hellebores under
the hazel, the yellowy-green Ballard hybrids with a maroon
splash at the eye, plus a fine blue mist of *Anemone blanda* under
the tree peony, which had just opened an abundance of crumpled yolk-yellow petals. Irises everywhere. Dying plants were
discarded and new plants took up their stations. A *Rosa rubrifolia*,
dead as a doornail. Great mats of lamium that had conquered
and pillaged the shadier border in the pond garden. A variegated euonymus, hideous anyway, and blocking two-thirds of
the path. Arbutus, sick, ceanothus, deceased. In their place I
planted more pinkish-green astrantia and *Geranium psilostemon*,
Verbena hastata 'Pink Spires' and a mass of tawny heleniums.
We got six goldfish, four orange and two black, and the garden
instantly felt more animated and alive. The tulips went over,
replaced by a wave of purple and white dame's rocket, aquilegia, purple drumstick alliums and the first glowing roses.

Morris thought everyone's environment could be and should be more beautiful. He believed that it was people's right to live in beautiful, unspoilt, unpolluted places, and he thought, like Ruskin, that beauty was not a luxury, and that luxurious and unnecessary things were actually unbeautiful, since beauty was so closely allied to necessity and nature. It was one of the founding principles of Morris and Co. that needful things had dignity and deserved to be made seriously and well. As his biographer Fiona MacCarthy puts it, the company's work was underpinned by 'the radical principles of products designed for the people by the people.' The fact that he found himself decorating the palaces of the wealthy, including the actual Throne Room of St James's Palace, was part of what shifted Morris towards his wholehearted embrace – bear hug, really – of socialism, his dramatic trans-formation from artisan to full-fledged Marxist activist.

In the essay 'How I Became a Socialist', written in 1894, he set out his beliefs very simply. 'Well, what I mean by Socialism is a condition of society in which there should be neither rich nor poor, neither master nor master's man, neither idle nor overworked, neither brain-sick brain workers, nor heart-sick hand workers, in a word, in which all men would be living in equality of condition, and would manage their affairs unwaste-fully, and with the full consciousness that harm to one would mean harm to all—the realization at last of the meaning of the word COMMONWEALTH'. The last word had not yet acquired its colonial associations, and instead represented a direct continuum from Winstanley's common treasury and Barmby's communism.

It's interesting that the kind of garden Morris created at Red House was such an open disavowal of the landscape gardens made by Capability Brown and his ilk, with their encoded message of a hierarchical social order as both natural and permanent. In this essay, and in much of his thinking of the time, Morris railed explicitly against what he called the Whig frame of mind. It was the same thing I'd found so disquieting while reading Horace Walpole, the same clapped-out and dangerous dream that continues to enthral politicians today: the faith in mechanical progress, the powerful belief in the virtues and benefits of industrial improvement, the adoration of profit as its own sunlit reward.

Morris was not simply being nostalgic here. In fact, the accusation of nostalgia can be seen as part of the same Whig mindset, the belief that humanity is moving perpetually upward in its attainments, and that to pause or reverse is automatically a negative and regressive act, with correspondingly devastating social and economic consequences. What Morris believed instead was that many of the decisions around progress had been wrong, that good and simple ways of doing things had been replaced by cheaper, quicker ones, which impoverished and made ugly the lives of many while making millionaires of a very few. He didn't hesitate to inculpate himself in this dynamic, as purveyor, employer or consumer, asking in one of his lectures the still unanswerable, still troubling and turbulent question: 'how can we bear to use, how can we enjoy something that has been a pain and a grief for the maker to make?'

For all that *News from Nowhere* is inflected by his love of the medieval world, he didn't want people cast back into a turnip-grubbing past. What he wanted was a future founded on equality, which cherished and held in common that most intricate and precious of all resources, the natural world. If only he'd been heeded in his warnings. If Morris saw in the industrialised, stratified and exploitative world of the Victorians 'sordid, aimless, ugly confusion . . . the dull squalor of civilisation', imagine what he'd make of the present day. Ecological catastrophe, species collapse, and still the unstoppable obsession with growth, the blind faith in technology as a get-out card. The metaverse, colonies on Mars, microplastics, coups carried out on Twitter: how Morris would have raged and grieved.

He regarded the creation of a new society as more important than art-making, his own included (he once wondered publicly if all the beautiful things he'd made amounted to little more than Louis XVI and his lock-making), but he also thought the choice between art and revolution was a false one, because art grew from 'thriving and unanxious life' and so required a great upheaval in the social order if it wasn't to become simply an outgrowth of capitalism, a pretty, pointless excrescence. Luxury and lassitude were not art as Morris understood it. If the artist did have work to do, it was this: to show the way ahead, and to create the nourishment and the desire to get there. What a great bulwark of faith this single sentence enacts.

It is the province of art to set the true ideal of a full and reasonable life before him, a life to which the perception and creation of beauty, the enjoyment of real pleasure that is, shall be felt to be as necessary to man as his daily bread, and that no man, and no set of men, can be deprived of this except by mere opposition, which should be resisted to the utmost.

The same enormous hopefulness animates *News from Nowhere*, a novel that is not so much a blueprint for a future society as an invitation to imagine what life could be like under changed priorities, without the fear or greed or precarity that capitalism engenders. How it could feel and smell, what its sensual attributes might be. How people might dress, or talk to one another. How relations between humans could change without the deforming effects of profit. There is no money in the England that Guest awakes into. People work because they want to, as gardeners do, out of the sheer love of making something. The capitalist system of alienated labour has melted into air.

By the beginning of June the garden had doubled in size, no kidding. It was floriferous and floppy, a sweetshop of foxgloves and lupins, Russell Hybrids in Quality Street maroons and purples that clashed happily with the soft new plumes of wisteria. The two 'Madame Alfred Carrière' roses that covered the front of the house were finally in flower, loose peachy rosettes that looked sublime until it rained. The two beds in front of the house had been covered in more weed membrane and gravel, punctuated by architectural clumps of indestructible sedge. Low-maintenance gardening. Earlier in the spring I'd yanked it all out, two days of labour, digging it over and slowly filling it with a profusion of cottage plants that the *florist* John Clare would have recognised, among them foxglove, fennel, hyssop, angelica and rose campion, along with a scattering of tagetes and cosmos seed, pretty summer annuals I thought he might have liked. Gardening for the street. Loose was my watchword, loose and lavish.

It's a sadness that Morris never saw Clare's work, which he would have loved, sharing as he did so many of its moods of rapture and teeth-grinding rage. He read Clare's older and more politically radical contemporary William Cobbett alongside Marx and Engels, Robert Owen and More's *Utopia*, during the period of feverish reading that formed the theoretical component of his socialist education. For the first year that he was a child at Woodford Hall he was only six miles from where Clare was then living in the High Beech asylum, so that they might plausibly have passed one another on their long rambles through Epping Forest. I like to imagine this, the boy in his armour riding on his pony past the little man who believed himself to be Lord Byron. But after Clare escaped in 1841, no further intersection was possible. Clare, who entered Northampton General Lunatic Asylum when Morris was seven and died there when he was thirty, had vanished from the world. The great majority of his writing went unpublished, untranscribed, unknown, waiting mute for history to rediscover it.

In one of these long-buried poems, 'The Wish', probably

written when he was still in his teens, Clare envisages his ideal house and garden (also his ideal companion: not a noisy pestering wife, but a modest servant who was witty and liked to read). This cottage strikes a distinctly Morris note, right down to its slates and British oak. It's modest and plain, with every needful detail precisely imagined and nothing of gaudy excess. The aesthetic is avowedly plain, simple, well-made and durable, the greatest luxury a fireside cupboard in the small parlour for keeping books. The garden, on the other hand, is a far more sensuous space. Its walls are planted all round with fruit trees: 'Peach and Pear in ruddy lustre glow'. The beds are a precise five foot, edged with parsley and harbouring no smothering weeds. The space is quartered by paths, with an arbour at one end planted with rose or jasmine or the sweet woodbine, honeysuckle, 'a compleat harbour both for shade or smell.'

Isn't this reminiscent of Red House? The plants in Clare's dream garden are fragrant too: roses, ranunculus, jonquil, lilies, chocolate-coloured scabious ('jocolately dusk'). He imagines climbers tumbling about the windows and the door. You can feel his excitement mounting. What about a pond?! A little pond full of waterlilies, set in a sunken garden walled with freestone. Something about this last space induces such an intensity of rapture that the whole lovely vision collapses in upon itself. The mood darkens and grows bitter. Clare has none of these things. He has been feasting on dreams again and his stomach growls. What he really wants is a life without labour, where he can tend his garden and read his books.

In *John Clare: A Biography*, Jonathan Bate describes this poem as an apprentice piece, in which Clare is mimicking the subject matter he thinks poetry ought to express, rather than the oddness and quiddity of his mature vision. I don't agree, not entirely. There's something of the lingering, longing accumulation of detail that feels authentic to its lack, and the shift in tone has an abruptness and rawness that doesn't match the smooth rendition of an exercise. Bate sees Clare's 'Wish' as a fantasy of switching classes, acquiring a life of gentlemanly ease and contemplation, the simple good life attested to by Horace and Virgil, Marvell and Voltaire, whose satirical novel *Candide* famously ends with the ambiguous injunction to cultivate one's own garden: *il faut cultiver notre jardin*. This life, he argues, would not have suited Clare, whose poetry arose out of a deep, perhaps bottomless well of loss and dispossession. 'Poverty', Clare answers back, 'has made a sad tool of me by times – and broken into that independence which is or ought to belong to every man by birthright.'

I was struck by what Bate said, though, because it inadvertently touched on one of the more radical aspects of Morris's utopia. There is a pervasive belief, in our own as well as in Victorian times, that people are driven to work and create out of a state of lack and need, and that if resources were shared more equally it would lead to widespread idleness and stasis. Strangely enough, this stasis often infects fictional utopias too, which having conjured a myriad of labour-saving devices and technological miracles struggle to imagine what the languorous citizenry might do with their days. There is an absence of tension, a slackening and slaking, all desire fulfilled, all problems solved.

The world of Morris's *Nowhere* is not like that. His people work because it gives them joy to do so. The person who voices this most clearly is Ellen, the beautiful grey-eyed girl who Guest first encounters lying on a sheepskin in a cottage very like the one Clare dreams of. 'In the past times,' she tells him:

> when those big houses of which grandfather speaks were so plenty, we *must* have lived in a cottage whether we had liked it or not; and the said cottage, instead of having in it everything we want, would have been bare and empty. We should not have got enough to eat; our clothes would have been ugly to look at, dirty and frowsy. You, grandfather, have done no hard work for years now, but wander about and read your books and have nothing to worry you; and as for me, I work hard when I like it, because I like it, and think it does me good.

Clare's fantasy also helps to clarify why gardens are so important in Morris's vision of the future. The cottage garden as Clare describes it is an idiosyncratic, intensely personal and creative space, and its presence in *Nowhere* makes it clear that the communism Morris is proposing doesn't mean everyone has to think or act the same. Sameness was anathema to Morris. What he liked was individuality amidst common purpose, each person as distinctive as flowers in a meadow. The same scene recurs in novel after novel, despite their varied time-frames: a craftsman or woman, engaged in the collaborative immensity of a creative project like a cathedral, takes pleasure in their own particular part, like Margaret in 'The Story of the Unknown Church', who carves the quatrefoils and the signs of the months.

Considering how much time he spent lecturing about the misery of factories and the suffering of their workers, the question remains as to why Morris didn't put his ideas about creativity into practice at his own factory, established in the old silk weavers' works at Merton Abbey. It was a romantic place to visit, with the fairy-tale, slightly ramshackle appearance Morris so loved: lawns rolling to the river, yards of coloured fabric drying in the meadow, an almond tree spilling its white blossom by the black-boarded sheds, a bed planted with larkspur and lilies that echoed the patterns of the chintzes. Some of the men rented plots in the old kitchen garden, but there were boys of twelve working in the unrelenting clatter and roar of the weaving sheds, bent double at the grinding, repetitive, mind-numbing work that elsewhere Morris spoke so passionately against.

His answer was uncompromising. You can't have socialism in a corner. He wasn't interested in setting up an ideal community, a utopia so isolated from the antagonistic world outside that it became a vulnerable island, like Robert Owen's New Lanark and New Harmony or Barmby's Morville Communitorium. All of these places inevitably failed. What Morris wanted was a total reform of the social order: not an exemplary factory but a Marxist revolution. In a letter to the editor of *The Standard*, he explained the starkness of the situation: 'we are but minute links in the immense chain of the most terrible organisation of competitive commerce, and . . . only the complete unriveting of that chain will really free us.' Morris and Co. was bread and butter work in the hope of jam tomorrow. Change would come via political activism, the exhausting, endless lectures and marches, committees and meetings, helped along by the dreams of the future he so carefully sowed.

His utopia was not an island. It was an Edenic republic without a God, almost certainly achieved by violence, since Morris was under no illusions about the tenacity with which the wealthy would hang on to their spoils. I want to spell out very carefully here what socialism meant to Morris, since it is such a perpetually misunderstood and distorted word. At minimum, it meant free education, free school meals, eight-hour days, decent housing, state ownership of banks and railways and above all the land itself, reversing the work of enclosure. At maximum, it meant a world where there was no concept of profit, surplus or waste, where the goal was not economic growth but the quality, the wealth of every person's life, as well as the ecosystem in which they lived.

Morris described his political awakening in 1883 in explicitly religious terms. It was a conversion, he always said, akin to crossing a river of fire. At the age of almost fifty he was joining a movement right at its birth, while it was still in the process of being drawn up. The historian and Morris biographer E. P. Thompson calculated that there were perhaps two hundred people in England who made the journey into socialism that year. Morris didn't know if a revolution would take place, or how the new society would function, but he was certain that for the first time in his unstintingly productive life he was engaged in real work.

Though he liked to present it as a Damascene enlightenment, a moment of flaming clarity, it was a slow decade of prods and pricks that finally woke him up. As he keeled through his forties, Morris had become frustrated with decorating houses, regarding his wealthy customers with what Thompson describes as 'increasing distaste.' He'd started using a factory in Leek in Staffordshire to dye his fabrics, and so had come a little closer to grasping the intimate miseries of the industrial poor. He was boilingly furious with the hypocrisies and war-mongering of the Gladstone government, too. It was because of these compound frustrations that this shy and powerfully tongue-tied man began making public statements and giving lectures on how work, art and society could be rethought.

Morris's political awakening was also fed by deeply buried emotional streams. At the age of six, his beloved daughter Jenny had developed epilepsy. It worsened in her teens and by the time of his conversion she'd become a permanent invalid. This is how we know so much about Morris's gardens, from the doting, closely stitched letters of his days that he sent to his daughter, as a way of keeping her connected to the world.

Then there was Janey, that most emblematic of the pre-Raphaelite beauties, gaunt, silent, supine on her couch. She was born Jane Burden, the daughter of an Oxford stableman and an illiterate laundress. Morris met her in the city when he returned after university to decorate the Oxford Union and she accepted his proposal, though she later said that she was never in love with him. He hauled her bodily into the middle classes, teaching her the cultivated arts of embroidery, Italian and the piano and plying her with books (it's said, perhaps unkindly, that the Morrises' friend George Bernard Shaw based the character of Eliza Doolittle in *Pygmalion* on Janey).

For many years, Janey had been caught up in an affair with Morris's close friend, the spiteful, saturnine Rossetti, who painted her dozens of times, dressing her up as Mariana in the moated grange and as Proserpine with her fateful pomegranate, trapped in Hades and gazing disconsolately into the sunlit upper world. It's likely, though not absolutely certain, that Morris took on Kelmscott Manor as a way of allowing Janey to see Rossetti in private, without risking scandal or divorce. Most Victorian men would not have behaved like this. Morris and Janey remained married, despite a second affair after Rossetti died with the poet and adventurer Wilfred Scawen Blunt. Who knows how couples work in private, what deficits they have and what bonds exist? The letters between Morris and Janey attest to more or less steady currents of affection and care, but it is certain that both experienced painful lack.

All of this was part of the fabric of Morris's conversion, its fluent, complex pattern. Humiliated, lonely, anxious for his daughter and deeply sad, he'd become less bumptious and thick-skinned. It was as if the divisions between himself and other people had started to wear away, as if the walls really were dissolving into the shifting tracery of leaves and petals that his patterns invoke. What separated him from the indigents who passed up and down in front of his Hammersmith house was nothing to do with inborn genius or talent, he realised, but simply the mad fortune of his birth.

It's this deep, infectious wistfulness for something better that animates his political and aesthetic visions. Both are saturated with longing for plenitude, for pleasure, for something that is glimpsed around the corner, that once existed and might come again. Is it love? Is it sex? Is it a new social order? What is the endlessly renewable Edenic cargo of which Morris dreams? I think what his gardens really stand for is fellowship: a humming, thrumming togetherness that transcends not only sexual desire but the human world itself. Call it a garden state: a cross-species ecology of astounding beauty and completeness, never static, always in motion, progressive and prolific. I want to live there, and the world won't survive much longer if we don't. It hasn't come to pass, this fertile revolution, and yet every time you look into a garden, the invitation is still there.

VI

BENTON APOLLO

O n the solstice I drove north, the whole midsummer day unfolding ahead of me, blue at its margins and gold at its centre. I crossed the Blyth by the church and took the back road past the strange Tudor farmhouse that I once visited with my friend Lauren, who was thinking about buying it until we got inside and saw what successive winter floods had done to the beautiful oak-framed rooms. The owners were camping in the ruins, though the word 'camping' hardly does justice to the style of their occupation. It was as if they were a pair of Royalist officers in hiding during the Civil War. On the table were the remnants of their supper – a candlestick, two glass rummers and a pair of pewter plates. There was a medieval stained-glass angel propped against the sink and at the top of a vertiginous ship's ladder a dim, winding succession of rooms that left a confused impression of tester beds and Chinese Chippendale. Someone had set out a David Hicks tablescape on a chest of drawers, arranging the horn drinking cups and ivory paperweights with military precision, though they were so thick with dust it might have been raining down since the house was first built.

Past the gun shop in Beccles, where I took a wrong turn and had to loop around by the Morrisons to find my way. Over the Waveney and then flush up through the Broads, oxeye daisies blinding in the verges, until the land dropped away beneath me and I could see Herringfleet and Somerleyton across the great flat expanse of the marshes, cut by the shining lines of dams and channels, and then the Waveney itself, which had swung around on its meandering course to the North Sea. I was early, as ever, and so I crossed the river again and parked by the round-towered church at Herringfleet, which Sebald or his pseudo self must have seen when he got off the train at Somerleyton Halt, at the very beginning of his walking tour in *Rings of Saturn*.

I've done enough church visiting not to push through the heavy door with expectations, but the stained glass at St Margaret's was far better than I'd hoped. It was a piecemeal collection of English and Continental fragments, put together in the 1830s but dating back to at least the fifteenth century. There were cherubs and phoenixes, saints and birds, an angel playing a lute and a parrot with a Latin scroll clasped in its beak. I saw several crucifixions and a sad man I thought must be Adam, girdled in spring leaves, collaged haphazardly alongside a tortoise, a mitre and a bunch of grapes.

All this is to say that I was already feeling light-headed when I turned the car, recrossed the river and reclimbed the scarp, parking by another little round-towered church. I'd come to see Mark Rumary's surviving partner, his companion after Derek Melville died, who I'd heard was a gardener of uncommon skill, though this encomium did nothing to prepare me for the paradise I found. My friend Howard once complained that most gardens are far too similar, unconsciously obeying rules of placement and content that were established centuries before, though there is really almost no limit to what can be done with plants. Even the so-called experiments rarely involve much more than a shake-up of the conventions concerning permissible shape and especially colour combinations. It's very rare to see something truly original, not a copy of a copy of a copy.

I wished he could have seen this garden. It was so unusual that at first I wasn't even sure what I was looking at. There were plants I didn't recognise, and even those that were familiar had been grown in ways I'd never imagined. It was a marvel of artifice and skill, one of the most sophisticated testaments I've ever seen to the gardening arts. You entered it by way of a brick flight of steps, passing through two piers capped with stone balls, which prepared the eye for an astonishingly architectural topiary garden.

It was as if someone had set out to illustrate ideal form by way of evergreens. There were lollipop wisteria and glob-ules of bay, free-standing beech arches and yew carved into pyramids and wedding cakes. A gravel path led through a maze of box parterres crammed with more specimens of topiary, many with variegated leaves, which enhanced the impression of unnaturalness, perhaps even decadence. What was this strange little tree, with its neatly plaited trunk? A lonicera, painstakingly trained into a standard, its tendrils arching down in plumes, barely recognisable as the honey-suckle that normally bulks untidily against a wall. The ceaseless variety of shapes and shades of green, interspersed with the fluent silver of a weeping pear and the flushed gold leaves of *Buxus sempervirens* 'Aureomarginata', was as deeply pleasing as a tapestry.

The garden rose up sharply to a dark wood dominated by two Scots pines. In the interstitial space were plants I'd never seen together. A colony of thrusting echiums, covered in the pale blue flowers so loved by bees. Was that a tree mallow beside them? Also enormous, also smothered in blooms, its leaves again variegated so that in places it was almost cream, growing in exuberant contrast to a cloud-pruned conifer I couldn't immediately identify, it was so far removed from the natural habit of any tree I knew.

Across the lane there was a gate into a second garden, which began with a formal grass walk past long colour-coded beds in shades of red, pink, blue and yellow. This looked far more like Mark's style. A bust I instantly coveted was at the end, set casually on a block of wood in a meadow of long, rain-battered grass. It was Antinous, lover of Hadrian, the most beautiful boy in the classical world. A white Chinese bridge led into another, larger meadow. When I stooped and looked the grass here was interwoven with vetch and lesser stitchwort and hundreds of flushed pink spotted orchids, growing wild in the damp ground. There was a nice eighteenth-century mound with a bench ahead, a match in period for the white bridge, but there was no time for sitting and contemplating the view. We were off again: passing through a dizzying repertoire of rooms that had been conjured up inside the wood.

A low box hedge like a stage set, encircling the sole player, a shapely magnolia tree, hung with a wind chime. Roses and foetid irises and martagon lilies, growing lush and unexpected under and into trees. Great elephant ears of gunnera and bamboo by a soupy brown pond, and beyond it a fanciful carved shed like a Russian dacha. You'd think you'd finally penetrated to ordinary woodland and then a yew made of perfectly balanced balls would loom between two cherries. The last station of the tour was a greenhouse used to propagate orchids, insulated with algal layers of bubble wrap and smelling sultry and damp. A pink clock on the wall suggested precision work. At eye-height orchid epiphytes had been lashed with green twine to logs wrapped in moss and suspended by hooks from a wire. This was gardening at an altitude I couldn't even fully comprehend.

I'd come to see Mark's photographs of our garden in its heyday, and after lunch we worked our way through five big leather-bound albums. They started in black and white in 1961, when Mark and his partner Derek Melville first bought the house. There it was, drab and unkempt and then freshly colour-washed and wired, the two 'Madame Alfred Carrière' roses just visible at the base. I was fascinated by the ways the building had transformed. Where was this scullery with a washtub and a range, and what had happened to the glass porch that opened from what was now our sitting room? Mark hadn't been exaggerating when he said how bare the garden was when he arrived, the mulberry and fastigiate yews standing ill at ease in a rough lawn muddled by brick paths.

Like Dorothy's entry into Oz, Mark's era was heralded by a plunge into technicolour. This was the garden in its floriferous pomp, a riot of supersaturation, the extraordinary height of the borders suggesting 1970s levels of fertiliser. It was very different to the zone of enchantment in which I'd found myself now: a little more conventional, a lot more luxuriant, just as romantic. I flicked on, fascinated, sometimes spotting the thriving plants whose corpses I'd dug out in the first autumn of our tenancy. Here was the ceanothus, spectacularly blue, and here was the tree anemone I'd read about and prayed had lived, covering the dining-room wall with its glamorous poached-egg flowers. I must plant another. The only photograph of Mark was of him reaching up smilingly to clutch a spray, dressed rather formally in a short-sleeved white shirt.

There were photos of the house's interior too, including one of Derek and a laughing woman I didn't recognise, clinking sherry glasses on what was obviously Christmas morning. After a while I realised something odd about these pictures. The room we were in now was an exact simulacrum of the ones I was looking at, every piece of furniture and ornamentation lovingly set in its proper place, right down to the polychromatic plates and jugs. Every room in the house, it turned out, had been furnished in the same way, recreating scenes established six decades earlier. It added to the dreamlike feeling of the day, to find that the dwelling place at the heart of this strange garden was suspended in time, encrusted with it, like an unstruck stage set preserved long after the run had finished.

It was Derek who'd collected the china. He was a composer and pianist who was obsessed with what the music of the Baroque period had sounded like on the instruments for which it was first written. He had a harpsichord and a clavichord made for him by Robert Goble in Oxford, played on surviving fortepianos in Vienna, Nuremberg and Budapest, and imported a grand piano from Paris, made in 1840 by his favourite composer Chopin's favourite piano-maker. The laurel in our garden was supposedly a cutting from the one on Chopin's grave, and he'd once written a biography of Chopin, too, though when I mentioned it to Ian, who has seemingly encyclopaedic knowledge of all books, he said it had bad reviews.

When Derek died in 1996 his ashes were scattered on the lawn ('we felt he would approve,' Mark wrote fondly, 'as he had always complained that I did not give the grass sufficient potash'). Later, Mark put together an elegant letterpress book in his memory, a collection of the Christmas carols he'd written once a year, describing in the foreword the lack of confidence and terror of public performance that had hindered Derek's career. After that, Mark seemed to have lived a more sociable, buoyant life. All kinds of people had told me about parties in the garden, and some of the stories were enviably wild.

I wanted a party too, and in August I got my wish. It was

the anniversary of our move and though the garden was nothing like I hoped it could become, it was the end of a full year of work. The watch and wait was over and soon I could plant in earnest, finally figuring out what sort of garden I would make from the lovely remnants of Mark's creation. Seeing those photographs had confirmed that I didn't want to recreate the technicolour vision of his youth, not that I had anything like the skill to do so. I wanted to make a wilder, softer space, that kept some of the secretive, lost feeling of the garden when I first saw it. And more than that I wanted to grow very old varieties, to play with opening up conduits of time by way of plants.

We had the party in the dog days of summer, just shy of Ian's birthday. I got out Mark's sign from previous garden parties, beautifully lettered by our neighbour, the artist John Craig: *DANGER FALLING MULBERRIES*. It felt good to open the house at last, to see the lawn fill up with friends, most of whom I hadn't seen in person in the last two years. On the invitation I'd pasted a few lines about the wedding party in *Cold Comfort Farm* that caught the festive mood I craved:

As the day emerged from the heat-haze, and the sky grew blue and sunny, the farm buzzed with energy like a hive. Phoebe, Letty, Jane and Susan were whisking syllabubs in the dairy; Micah carried the pails of ice, in which stood the champagne, down into the darkest and coolest corner of the cellar . . . Reuben was filling with water the dozens of jam jars and vases in which flowers were to be arranged . . . The air smelled sweet of cherry pie . . . Flora took a last look round, and was utterly satisfied.

I was just as satisfied. People of all ages were bubbling and spilling over, flopping on blankets and benches, their arms entwined. There was a makeshift bar, with buckets of champagne on ice, and great armfuls of flowers from Paula's roadside stall at the top of the hill. Ian had produced a feast, singing to himself as he went. No cherry pie, but the kitchen table was laden with sausages and hams, buttered rolls, warm madeleines and chocolate cake. For weeks afterwards, I kept finding corks in the borders, mementos of long overdue revelry.

But something about the combination of that subtle garden on the hill and the house that was so doleful an announcement of loss stayed with me, entangling with a phrase Mark's executor had used when we first spoke on the phone: *He was gay when it wasn't good to be gay.* Why did anyone make gardens? Why did I?

When we left Buckinghamshire we ended up in a commuter village on the south coast. It's not hard to articulate what was so wrong with those years, the second, unchildish half of my childhood. We were out of place, definitively strange. Two adult women, two small girls did not constitute a family in those relentlessly homophobic years. The house we landed in was made dangerous by alcohol, periodically illuminated by flashing blue lights. I remember it with a clenched feeling: freezing at the sound of a raised voice, trying to soothe or pacify, though it can hardly have been my job. The angry feeling: living alongside an adult's perpetual soreness, rages and tirades, their obsession with neatness and order, which was so powerful that many years later, when I kicked over a glass of red wine while getting off a sofa on Christmas Eve, in the late night company of my sister and her boyfriend, I was so paralysed with shock and terror that I spent hours scrubbing it and sobbing, and didn't sleep all night until I could confess. I was thirty-five.

This annihilating tidiness infected the garden too. I had fantasies about making my own garden among the shrubbery, as the young William Morris did at Woodford Hall. I loved plants, plotted in particular over a 1985 book called *The Wild Garden* by Violet Stephenson, but there was nothing transporting or enclosing about this new garden. It was implacably sterile, the tightly mown lawn planted with stripling oaks, two of which were torn up by the roots in the Great Storm of 1987.

We ran away in the end, my mother, my sister and I, a real John Clare flit. The final garden of my time as a child was on a brand-new estate, a maze of dead ends built so that every house was overlooked by dozens of not quite identical others. I've written about it before and been criticised for complaining, when I was lucky to have a house at all. That's true, of course it is, but the estate felt particularly exposing during that period in the early 1990s when being gay was in the public eye both a fatal sickness and a crime. My mother magicked a beautiful garden out of thin topsoil over builders' rubble, but despite the Russian vine that swallowed the back fence it was hardly the deep cover we required. We stuck out like a sore thumb.

Over the years, I've had many different ways of thinking about this past, some more bitter than others, but I find myself wondering now if the homophobia of the period was as much to blame as the drinking; if, in fact, it was the objective correlative that drove all the other catastrophes. To hold that kind of secret, and to understand by holding it that there was something that needed to be hidden, and consequences if – or more realistically when – it wasn't: it was like living inside a pressure cooker, never mind a closet. I felt that, and in those days I was only the witness or proxy, not the participant, so how much more enraging, frightening, omnipresent must it have felt to the adults in the house? The need to stay concealed, the corrosive sense of being constantly judged, constantly in danger, at the margins, secret, unable to trust; it was really not so very far from being a spy. Mark, I remembered then, always referred to Derek in print as his *friend*.

We all travel into adulthood bearing particular burdens, some of which are necessarily personal, individual, idiosyncratic, and some of which are more properly categorised as political, to do with the kind of history doled out to the people who shared our circumstances. I think there are many ways that adults handle and manage the sometimes radioactive material of their own past, and that I'm not alone in finding solace in the act of making gardens. The particular childhood I had left me with a craving, a permanent, nagging need for a place that was safe, wild, messy, bountiful and above all private. I wanted a home, absolutely, but it was a garden that I needed.

Realising this has made me think differently about what people are doing in their gardens, aside from simply making pictures or creating a space to spend a Sunday afternoon. Over the last year of work, I'd noticed in myself a constant toggling back and forth between a need for control – evict the lamium! clarify the borders! – and a desire for plenitude. As family crises continued to break, as I sat through meeting after meeting with lawyers over the ownership of my father's house, I'd become aware of the role the garden played in how I managed feelings. I'm not just talking about mowing the lawn to work off a bad mood, or weeding a bed in an access of grief. I'm talking about how we – and I mean here the 'we' of people who have experienced some kind of trauma, which I think is a very large category indeed – how *we* manage the serious burden that we carry, how we contain it, what spent fuel pool or dry cask storage we improvise for material that is even now still leaking its lethal isotopes.

If I had a manual for constructing this kind of device it was *Modern Nature*, Derek Jarman's memoir-cum-diary about making a garden on the shingle beach at Dungeness, after he had been diagnosed HIV-positive. It arrived in my life in 1991, bang on time, emerging out of the same desolating set of circumstances – Thatcherism, now congealing into the cricket and warm beer nostalgia of John Major; the Aids crisis, ten years without a cure or even reliable treatment; and Section 28, with its spiteful delegitimisation of the gay family. That book conducted me, as it did thousands of others, through a secret network of tunnels into another world altogether, luminous as Dorothy's Oz, the first film Jarman ever saw.

It was Jarman the time-traveller who attracted me then: the plantsman whose love of plants was partially due to their capacity for bringing the past eddying up to the present day. His tone gossipy, knowledgeable, donnish. '*Rosa mundi*, rose of the world, with its crimson and blush striped flowers, an old sport from the apothecary's *Rose officionalis* the rose of Provins. It was brought back by a twelfth century crusader and immortalised by Guillaume de Lorris in his poem the *Roman de La Rose*.' Never mind that the botany wasn't always quite solid (did he mean *Rosa gallica* 'Officinalis'?), or that I didn't understand half his references in those days.

The early sections in particular are thicketed with Renaissance and medieval plant lore. Here's the utopian Thomas More on rosemary: 'I let it run all over my garden walls, not because bees love it but because it is the herb sacred to remembrance and therefore to friendship, whence a sprig of it hath a dumb language.' How I wanted to be fluent in that dumb language, to read it as freely as Jarman could. At the end of *Jubilee*, his Elizabeth I says something similar, standing amid the rock-pools at Dancing Ledge in Dorset in her white brocade dress: 'the codes and counter-codes, the secret language of flowers'. Hieratic, mystifying, immensely old and changeable, endlessly renewing itself.

It was by way of *Modern Nature* that I first encountered Gerard, Galen, Culpeper, who I would spend my twenties studying, though the spell invoked by Jarman was almost worn away by the poultice of new language we had to use, *evidence-based, pharmacognosy, cytochrome P450*. It didn't return until I was back in the garden and the sentences came seeping up like groundwater. *I borage bring courage. The fields are garnished and overspread with these wild poppies.*

I took my copy up again after our party, for what might have been the twentieth time. At each new reading I am older, hold more in my own head – come closer, in fact, to Jarman's age at the time of writing, which I was startled to realise was only forty-six at the beginning of the diary – so that it continues to open up, containing things I'd failed to notice or understand on earlier passes. I knew, for example, that it was here I'd learned that 'paradise' is the Persian for 'garden', but I was surprised by how foundational the subject was. Reading under the magnolia, which cast rippling shadows across the pages as they turned, it seemed above all a meditation on paradise lost and Eden regained.

The first Eden was Italy. Jarman moved there in 1946, at the age of four, a staging post in a peripatetic military child-hood that also included Karachi in Pakistan. His father Lance, known as Mike, was a commandant of the Rome airfield and a witness in the Venice war trials. Derek's earliest memories were of wandering unsupervised in a sequence of melancholy and enchanted gardens. A villa on Lake Maggiore, where the grounds ran along the water for a full mile: a humming green world of overgrown camellia avenues and fallen statues in sweet chestnut woods, guarded by its sinister chatelaine, the wife of a disgraced Fascist. Next came an apartment in the Borghese Gardens in Rome, requisitioned from a relative of Mussolini. In the park there was a water clock set in a brackish pool and a mysterious Egyptian gateway in the shape of twin pylons. The city was half-empty in that first year after the war, populated by wounded soldiers and begging children, the Via Appia still a country lane wreathed in roses. This latter image made such an impression that as an adult Jarman listed it as one of seven entries under the heading 'curriculum vitae Art History self portrait romance'.

Imagine disembarking in England after that. Trailing roses and scurrying green lizards replaced by army camps and dull suburban gardens, rationing, grey skies, wet Sunday afternoons, lingering sadness, and then a second rupture at the age of eight, when he joined the 'bedraggled little boys in formless grey suits' at prep school. Hordle was a world of perpetual discipline and lack, a premature training in doing without one's most intimate and heartfelt needs: for touch, love, privacy; even sufficient food. Unsurprisingly, Jarman foundered there, struggling with the constant sports (his biographer Tony Peake lists running, boxing, swimming, sailing, rugby, cricket), bottom of class at his lessons, flourishing only in the garden plots the boys were allowed to tend, where he cultivated prize-winning rows of love-in-the-mist, cornflower and alyssum, losing himself in hours of floral contemplation.

Home was not much better. The family moved with each new posting, so that sometimes he returned from school to a beautiful Elizabethan manor and sometimes to a house 'blotched' with camouflage and surrounded by barbed wire. His father failed to shake off the commanding officer when he stepped through the front door. He drilled his family, insisting on a perfect discipline and order that extended outward to his arid gardens. He was, Jarman wrote, 'happiest with the axe', ruthlessly hacking back or dousing with weed-killer any plant that had become too luxuriant, the same brute mode of cultivation he applied to his elegant, excitable son, who was belted for any minor infraction, like saying *pardon* or declining to eat his greens.

At twelve, Derek graduated to public school. He was sent to Canford in Dorset, a striking architectural jumble of a building, its immodest tower designed by Sir Charles Barry, the architect responsible for the final flourishes at Shrubland Hall. It was set amidst the neglected splendours of an old iron-master's park; out of bounds to boys, of course. In *Modern Nature*, Jarman describes this place in decidedly similar terms to those used by Sebald on Ditchingham (the two men were almost exact contemporaries, Sebald saturnine and melancholic to Jarman's irrepressible sanguinity). This landscape, he thought, was a kind of conjuring trick, a malign alchemy, not at all well-intentioned: 'the park and idyllic acres so carefully constructed to conceal the origins of wealth, to cloak with gentility the dark satanic furnaces of South Wales.'

Not all seeming paradises could be trusted, was the lesson here. Many had evil origins, or were places of oppression, bullying and cruelty. *Paradise Perverted*. The school was a crib of Empire, with its training rites of fagging and corporal punishment, its perpetual observation and inspection, its sneaks and spies. Isolated and under threat, he refused to capitulate, despite briefly contemplating suicide in the Stour. Instead, he withdrew, then took off sideways, a survival technique he'd utilise all through his life. He set up camp in the art shack, where he threw himself bodily into becoming a painter, creating his own subversive curriculum. When he could, he carried out small acts of defiance. As a member of the obligatory Combined Cadet Force at Canford (a youth version of the armed forces), he was once spotted by a friend drilling boys in field work. As the observer drew closer, he was amused to realise that Jarman had taken the commission literally. He was instructing them in the Latin names of all the nearby trees.

It left a legacy, all this. Things that couldn't be said, feelings that couldn't be felt, a deficit of love and freedom, above all an agonising lack of touch. My father had a similar kind of education and once told me that, sent away from home at the age of eight, he'd slept with one arm stretched up above his head, so that he could pretend a warm body was beside him. Watched all the time, punished all the time, so that inevitably you became your own policeman, set to watch for any breech.

There had been one single Edenic interlude. At Hordle Jarman encountered sexuality in the blissful form of another boy. They cuddled in bed and licked each other all over in a secret bower of violets, 'such was our happy garden state.' It was a heavenly refuge from the bells and bullying and virtual starvation, until inevitably they were caught one night and dragged apart; 'like two dogs', he once said. Shame had entered the garden, making him even more painfully estranged from his own body. He didn't find it again, the spell of absorption and delight, time stopping on its heels, until adulthood, when he finally managed to divest himself of the misery and self-loathing he'd been forced to ingest, and admit that what he really wanted was to hold, kiss, fuck and love another boy. From then on he was O U T, a gleeful celebrant in the act of queer sex. In his first memoir, *Dancing Ledge*, he wrote that after his first night in bed with someone it was as if a great tide of self-hatred receded, leaving him shaken, 'like an empty shell from which a dark death's-head moth had escaped.'

In *Modern Nature*, there are two Edens: the garden he was making amidst the shingle of Dungeness and the wild bower of Hampstead Heath, where he went to cruise, getting a nocturnal taxi to Jack Straw's Castle from his flat in Phoenix House on Charing Cross Road. 'All the Cains and Abels you could wish for are out on a hot night, the may blossom scents the night air and the bushes glimmer like a phosphorescent counterpane in the indigo sky . . . No one who comes here need leave without an orgasm.' He writes about the Heath with such excitement,

such utopian intensity, as a place where all the hateful boundaries between race and class and wealth have temporarily
dissolved, where the city surrenders to the erotic reversals of a
midsummer night's dream. Perhaps it didn't occur to him that
it was still segregated by gender, though the least comfortable
aspect of reading Jarman for me these days is the faint rustle of
misogyny around his longing for a homotopia, an uncontamin-
ated world of men.

Aids despoiled his garden of earthly desire, coming as a
second, intolerable betrayal, a punishment of those who'd
chosen to make, in the phrase of his youth, love not war. He
still went to the Heath, but these days it was more often to
watch or to talk, counselling frightened young men or thrilling
himself with the antics of a skinhead or off-duty squaddie,
glimpsed in the light of bonfires beneath the great oaks. Hence
the book's urgency, written under a death sentence, time
running out: 'before I finish I intend to celebrate our corner
of Paradise, the part of the garden the Lord forgot to mention.'

But if his paradise was sex, it was also the dream of fusing
art and life, so that the one rose seamlessly out of the other, a
fantasy that stems straight from William Morris. Jarman lists
Morris in a collection of artists he thought were driven by the
same perhaps impossible dream, among them William Blake,
playing at Adam and Eve with his wife in his garden at Lambeth,
and the gay utopian socialist Edward Carpenter, who established a cross-class vegetative Eden in a smallholding in
Derbyshire. Walt Whitman, Eric Gill, John Berger, 'all of them
looking backward over their shoulders – to a Paradise on earth.

And all of them at odds with the world around them.' The gardener was the lucky one who'd found the key to eternity, the prize they all sought. Time stopped in a garden. Even as an unhappy little boy he'd known that. At Prospect Cottage he was recovering what school had annihilated and film-making had so often failed to become: a dreaming refuge, an other-worldly home, where the hours are not timetabled but drip like honey from the spoon.

The funny thing was that he hadn't gone to Dungeness to make a garden. The shingle beach was too inhospitable, and there were constant winds, the easterlies in particular laden with killing salt. He began with stones, not plants: the grey flints he called dragon's teeth, revealed by the tide on morning walks. Then a dog rose, staked with a 'curious' piece of driftwood, followed by dozens and then hundreds more plants, some drawn from his childhood memories and some from the medieval herbals he read on stormy evenings, many of them grown thriftily from cuttings. Others – white campion, restharrow, dog rose – were salvaged from the Ness itself, John Clare style, so that the garden continually undid the distinction between cultivated and wild that a garden is meant to proclaim.

In fact, it's hard to think of a garden that is more a product of its place, a relaxed embrace of its own habitat and locale. It's an artful rearrangement of what was already abundantly present: gorse and sea kale, driftwood and flint, the latter pains-takingly lugged home in a plumber's bag and set into beguiling new designs. From this austere skeleton the garden emerges in luxuriant detail, a time-lapse tide of colour – 'Mrs Sinkins' pinks, wallflower, rue, santolina, cistus, poppies – rising to swallow and ebbing to reveal sculptures made of rusty anchors and the corkscrews of anti-tank fencing left over from the Second World War, capped with bones and necklaces of the holey stones he loved. He saw it from the beginning as 'a therapy and a pharmacopoeia'. It was a place of total absorption, and it was this capacity to slow or stop time, as much as its wild and sportive beauty, that made it such a paradise-haunted place.

The garden is most often spoken of these days as a response to Jarman's diagnosis and to the Aids crisis, an upsurge of wild creativity in the face of almost incomprehensible loss and devastation. This is undoubtedly right, but after going back to *Modern Nature*, I felt certain it also had roots in the childhood needs that had gone so badly unfulfilled, which he was teasing out on paper at the same time that he was making the garden. The longing for beauty. The unspoken, unspeakable anger and fear, tamped down and still festering somewhere. In response, he induced a desert to flower, providing himself with a lush wilderness, a sensual, anarchic space, the opposite of the malevolent austerities in which he'd been raised. It's such remedial work, making a complex, living structure, every element by hand. It seemed to me that the garden was not so dissimilar in its function to the way he'd described the art shack at Canford: 'a fortress against another reality, a defence against an everyday existence that was awry.' Back then he filled its moat with painted flowers, because he no longer had anywhere to grow them. Now they were real.

I've visited the garden at Prospect Cottage many times over the years, though I've only once gone inside the house. The first trip was the year after Jarman died, on 19 February 1994 in St Bart's Hospital. I came with my sister and my father, and to our confusion we couldn't make out where the garden was. We were used to seeing it in Howard Sooley's beautifully framed, super-saturated pictures, and hadn't quite grasped how open and unguardedly it lay in the shingle desert of the Ness. And then Derek's partner Keith Collins, nicknamed HB, came out of the house with a basket and began to hang up his laundry. The sense of trespass was too much. We backed away without really looking at a single plant.

Years later I met Keith, when I was writing the introduction to the reissue of *Modern Nature*, and also spent a long strange evening on the phone to him, during which he told me the story of his life. Every time I thought I'd trespassed on his time enough he began another story. When we organised a reading at a London book-shop he turned up a few days early and offered to fill the window with Jarman's gardening tools. Still as superlatively beautiful as he was as a lad, his dark hair flowing down his back. Three months later he was dead, killed by a brain tumour out of nowhere. Somewhere I have a photo of him cuddling a cheetah, his full lips and sea-green eyes. I don't know where I found it, floating amidst the jetsam of the internet. That night in London he wouldn't join us on stage but spoke from the audience about the threats facing the NHS, which had done so much to care for Derek.

The feeling that the garden was a shrine, a reliquary for ghosts, was even stronger on my final visit. I went with Amanda Wilkinson, who looks after Jarman's paintings, on a bright December day in 2019, and she let me into the house and left me there alone. I thought I'd be overwhelmed with delight, but it was drab and sad and I wanted to escape. Dried up tubes of paint, old address books, a Polaroid pinned on the kitchen shelf alongside ancient tins of food. Outside, the flint dolmen guarded no one. The garden was tended, but it was no longer serving the same pressing private function. This place had been the site of so much energy and enterprise, such restless motion, and now it was only a shell. What I'd loved was the reckless aliveness, preserved intact in the books and films, but not here, in this magnified, echoing absence. A garden dies with its owner, I was beginning to see that. It can be re-animated in a new form, absolutely, but the tutelary spirit who'd made all this, the old magician on his driftwood throne: he'd left.

At the end of August I mowed the little meadow, raking it bare to sow the flat brown seeds of yellow rattle, which is parasitic on grass, helping to shift the balance of the lawn back to wild flowers. On the hill above Rookery Farm they were bringing in the wheat, the combines working back and forth in the shining field. We gorged on yellow bullaces in the hedgerows, so ripe they burst when you bit into the skin. A new garden was emerging in my mind, rising out of the bones of the old. I could almost see it, I thought about it so much. Ian was going to turn the coach house into a library, and I wanted to make the derelict space where the wedding had been into a library garden, as a gift for him.

I'd dreamed it up almost as soon as we arrived. The first mention in my diary was from a full year back. 'Planning Italianate formal garden', I wrote, 'with stone flagged pond, gravel, and then deep borders of shade-loving plants, roses on the walls, the canopy subtly opened up and reduced.' Beneath this entry was a rough sketch, with a rectangle inside a circle inside a square. A page later, I went back to the subject. 'The pond should be deep, formal, clear, very quiet restrained space, quiet plants, fragrant, calm, two urns, wood anemones and woodruff, pale, still, a seat by the coach house wall, white roses.'

At the moment the hogweed was head-high by Matt's rambling rose, coiling along the back wall. The dead and dying cherries had gone but there was still a dense shrubbery in the bed against the yew hedge. I'd abandoned the idea of gravel in favour of a flagstone circle in the middle, with the deep rectangular pond at its centre, reached by existing paths on either side. The first of these was brick, passing under the broken archway with its festoon of ivy and along the north wall, entering what was currently still a ragged circle of tarpaulin between two flourishing box plants I planned to clip into cubes. The second path entered by way of a softer, darker arch made by the two Irish yews, passing beside the wall of the coach house, where one day there'd be French windows to fill the library with light.

There were many plants I wanted to keep. The Chopin laurel and the medlar, of course, its shape like a tormented umbrella. A hoheria that was smothered in tiny fragrant flowers that fell across the stones like dust. A choisya, that much-maligned car-park plant, whose leaves smell of blackcurrant and whose pearl-white flowers will come twice in a warm year. The border at the back in particular was full of treasures. A white *Rosa rugosa* 'Blanc Double de Coubert'. Skimmia, male and female. The white peony I'd found beneath a Japanese holly fern, plus the colony of shuttlecock ferns that made the garden such a joy in early May, when the medlar blossom was out and it became an enclosed chamber dressed in linen-white and sap-green.

I wanted to embellish this cloth with a fine embroidery of woodland plants: sweet violet and wood anemone, rusty foxglove and martagon lily. It needed more roses too, and in my email drafts folder I kept a running list of possibilities, all of them old or wild varieties. *Rosa* x *cantabrigiensis*, for sure, with its ferny leaves and masses of primrose-yellow single flowers. It was bred from two wild parents at Cambridge Botanic Gardens in 1931, where many years later Ian used to walk with his small sons. Then perhaps another *Rosa rugosa*, pink this time. I was inclining towards 'Roseraie de l'Haÿ', which produces heady clouds of scent. The Dunwich rose, for the drowned medieval village where we swam? Or maybe *Rosa complicata*, with its clear pink flowers, loved by bees?

I imagined tree peonies in soft yellows and dark wine-reds. A quince and a crab apple where the shrubbery was now, underplanted with woodruff and astrantia, a foam of Elizabethan plants. Delving about on the Rare Plant website, a dangerous occupation, I found fritillaries called 'Eros' and 'Saturn'. I could introduce them under the medlar as a nod to Jarman and Sebald, perhaps. Old irises, certainly, and here I began another list. *Iris pallida*, with its soft bluish flowers. *Iris florentina*, whose roots, Gerard says, are 'exceeding sweet of smell' and were known in shops of the time as orris or ireos, 'whereof sweet waters, sweet powders, and such like are made.'

The language of irises is so bewitching. 'This Iris hath his flower of a bleak white colour declining to yellowness.' What about 'Neglecta', from 1835, or 'Mrs George Darwin', bred in 1895, its white flowers marked with heavy gold reticulation at the hafts changing to violet on the upper falls? Though if I wanted perfection of name and form it really should be one of the Benton irises I'd choose: perhaps 'Benton Olive' or 'Benton Apollo'.

According to my old iris bible, *The Tall Bearded Iris* by Nicholas Moore, Benton irises are 'notable not only for their substance but also for the characteristic delicacy and artistry of their markings.' They were bred by the painter-plantsman Cedric Morris at Benton End, a Tudor house on the outskirts of the Suffolk village of Hadleigh, not so very far from us. Morris came to Benton in the summer of 1939, with his lover and companion, the artist Arthur Lett-Haines, known invariably as Lett. The pair had been running a small and decidedly unorthodox art school, the East Anglian School of Painting and

Drawing, in nearby Dedham. Earlier that summer it had burned down, and they were in search of a new premises, large enough to accommodate students and where Morris could make a garden.

Both men were striking. Lett was six feet tall with a scarred shaved head, while Morris dressed like an elegant tramp in rumpled corduroy, with a dandyish red kerchief knotted around his neck, a pipe-smoking caricature of what he was: a socialist baronet. Together they built a paradise at Benton End, an open-minded enclave of hedonism and hard work. As the writer Ronald Blythe, a frequent visitor as a young man, explained: 'Lett and Cedric were open about their homosexuality at a time when it was illegal to have such a relationship and they also conducted a fight against the philistinism of the day . . . The greatest crime at Benton End was to be boring!'

Over the years a mass of students passed through this charmed enclave, among them Lucian Freud and Maggi Hambling – the person, I remembered now, who had given Jarman the title of his book, describing his new set of pre-occupations as *modern nature*. Gardeners and artists visited – Constance Spry, John Nash, Vita Sackville-West, Francis Bacon – but alongside these famous names were many lost or disaffected young people, often gay, who were drawn by the sense that something alive and different was happening here. They were given jobs and nicknames – The Bird, The Little Prince, The Royal Bum – and encouraged to paint or garden, according to their inclinations.

The atmosphere of homophobia had become far more intense in Britain after the war. When the Conservative Home Secretary Sir David Maxwell Fyfe was appointed in 1951, he ordered a drive against what was then termed 'male vice', with the infamous promise to 'rid England of this plague' (though much quoted, it is impossible to find a source for Fyfe actually using these words). Fyfe drastically increased policing, deploying undercover officers to pose enticingly in parks, public lavatories and other cruising grounds ('the prettiest ones,' Jarman recalled in *Modern Nature*. 'They had hard-ons but didn't come. Just arrested you.') In December 1953, Fyfe reported to the House of Commons that in the previous year there had been 5,443 arrests for so called 'unnatural offences' and around 600 imprisonments, an enormous increase on previous decades.

'Homosexuals', he added, 'in general are exhibitionists and proselytisers and are a danger to others, especially the young, and as long as I hold the office of Home Secretary I shall give no countenance to the view that they should not be prevented from being such a danger.' The victims of his post-war witch hunt included the computer scientist and war hero Alan Turing, who was prosecuted in 1952, choosing treatment with the female hormone oestrogen as an alternative to prison, and Lord Montagu, whose trial provoked a scandal. Decades later, Montagu remembered a friend joking bleakly that 'the skies over Chelsea were black with people burning their love letters.' Suicides were legion; blackmail pervasive. In 1954, Turing killed himself with the poison cyanide.

Benton End was a refuge from all that, a place where people could be themselves, relinquishing some of the pervasive fear and secrecy of those years. Lett and Morris befriended gay men recently released from prison, while their housekeeper Millie Hayes was a former model and student to whom they'd given a home after she emerged from two decades incarcerated in an asylum with post-natal depression. The garden walls protected them from the hostile, prying, often dangerous world outside, permitting the creation of a fertile counter-state. In *Benton End Remembered*, which gathers together reminiscences from dozens of these people, one of the strongest notes is gratitude. 'In one way or another', said Blythe, 'all of us "flowered" at Benton End.'

The conversation in the squalid kitchen was so bohemian, so cosmopolitan and wicked that one visitor described it as 'almost like stepping into a novel.' Lett claimed he'd taught his friend Elizabeth David to cook. He presided over the filthy Aga, wine glass in hand, complaining vigorously. 'The buggers don't know what they're eating,' Maggi Hambling once heard him say. 'They might as well have ham sandwiches.' Instead, he produced elaborate fish stews and fruit jellies even in the years of rationing. 'Chinese fare and coffee-chocolate blancmange for lunch,' one student wrote wonderingly in 1945. 'Lett certainly can cook!'

Morris was a gardener to his bones, while Lett could rarely be persuaded to step outside, preferring to spend his afternoons in bed, recuperating with champagne and a tin of sandwiches from the arduous work of keeping house and school afloat.

After the war, when he grew mainly vegetables for the kitchen, Morris had transformed the three-acre walled garden into a plantsman's paradise, filled with rarities he collected each winter during painting trips to Spain and Portugal, Cyprus, Lebanon, Libya and St Helena.

The garden writer Beth Chatto made the first of three decades of visits in the early 1950s, when she was a young farmer's wife just beginning to become interested in plants. Her friend Nigel Scott had procured them an invitation to tea. She recalled being led into a vaulted pink barn of a room, the walls crammed with paintings of flowers and birds, the long oak table crowded with people, every surface covered in a bric-a-brac of china and drying plants. A Suffolk policeman's daughter, it was her first glimpse of la vie de bohème. After tea they were taken into a garden that was unlike anything she'd ever seen before. There was none of the familiar structure of beds-plus-lawn, but instead 'a bewildering, mind-stretching, eye-widening canvas of colour, textures and shapes, created primarily with bulbous and herbaceous plants'. She wandered through it that first afternoon 'like a child in a sweetshop, wanting everything I saw'. Nigel, even more covetous when it came to plants, promptly moved in and became Morris's lover and accomplice in the garden.

What might they have seen that afternoon? Species roses and hellebores, great drifts of fritillaries in lemon and plum, with speckled or netted interiors. Alliums, giant yuccas and abundant poppies in a strange, shimmering greyish-mauve like a rain-cloud. Alpine strawberries, double primroses, tree peonies and pear trees hung with curds of blossom. On and on the dreamy lists go: tiny upright widow irises, in their crepuscular green and black habits, muscari, scilla and yellow asphodel, which is said to carpet Hades and provide food for the dead. One wild rose in particular had grown so large that it formed a cave you could crawl inside. Morris continued gardening well into his eighties, and could often be found after lunch curled like an old cat among the flowers, fast asleep, while the bees worked over his head.

That fecund version of the garden is long gone now. After the two men died, Lett in 1978 and Morris in 1982, the house went into private ownership and a strict mowing regime turned Morris's jewelled meadow into an ordinary lawn. He'd sensibly dispersed the majority of his plants by way of a plant executor, his friend and neighbour Jenny Robinson, who in turn entrusted many of them to Beth Chatto. Chatto too died in 2018, but plenty of those plants are still being propagated in her nursery outside Colchester, where Jarman also sometimes shopped, so that in a way Benton End is always being disseminated, seeding itself further and further abroad. In May 2020 it was announced that the house, like Prospect Cottage, was being rescued, with a restoration programme established under the stewardship of the Garden Museum. Once the spring mowing stopped, it turned out that many of Morris's bulbs had survived, the *Fritillaria pyrenaica* with their subtle chestnut and sulphur striations still gleaming like bells in the long grass.

These revenants aside, the exuberant disarray of the garden's glory years is best preserved in Morris's paintings. He was above all a colourist, creating sensuous, almost hallucinatory arrangements of vegetative forms, composed in clean, shining pats of colour. He loved to bring tones into seductive, unsettling intimacy: ripe orange alongside imperial purple, the sharp yellow of crown imperials against the acidic green of the Mediterranean spurge, *Euphorbia characias* subsp. *wulfenii*, ubiquitous now but in those days very unusual. 'It is not only the intensity of his colours that tells,' Richard Morpeth, the curator of the 1984 Tate retrospective once told Ronald Blythe, 'but above all the originality and strange beauty of the relationships between them that he establishes.' Often he painted in the *plein air* of the garden, working in the unusual manner of starting at the top left corner and proceeding calmly in rows across the canvas until he had reached the bottom right, like somebody knitting. He treated his plant subjects like human sitters, conveying the camp idiosyncrasies of their character, their shyness or flamboyance. Years later Blythe commented that he was not dissimilar to John Clare in that he made friends with flowers.

My favourite of these paintings is *Benton Blue Tit*, from 1965. It shows the tall red house at Benton End almost eaten up by a miscellany of June blooms: bearded and Siberian irises, the fubsy grey tips of mullein, the wiry bells of *Nectaroscordum siculum*, allium heads gone to knobbly green seed. What is most powerfully conveyed is the particular feeling of each flower, its essence, if you will. He subtly exaggerates their most distinctive attributes: the thrusting pollen-heavy orange stamen of a martagon lily; the peculiar enamelware blue of an echeveria. The clematis in particular is huge, tumbling into frame, superabundant and luxurious, straight from Milton's Eden.

The iris was Morris's most prized plant. After his paintings drifted out of fashion amid the new abstraction and kitchen sink realism of the 1950s, he was far better known as a breeder of plicata irises than as an artist. Each year he raised a thousand seedlings of his own crosses, breeding for subtlety of colour and perfection of form. The results were shown at annual Iris Days in the garden, with tea and rock cakes at four o'clock. The petals of plicata irises are lined with fine veins of colour, and Morris's gift as a colourist brought innovations of exquisite understatement and sophistication. I'd been reading recently about the colours of Elizabeth I's dresses, and their names recalled the off-kilter subtlety of Morris's palette. Watchet blue. Ash, dove, carnation, russet. Bee colour, hair colour, drake's colour, clay, horse-flesh, lady's blush, maiden's blush. Isabella, murrey, partridge and straw. Cloth of silver, cloth of gold, sea-green, peach and violet. Imagine rank after rank of those petals and falls.

At the height of his breeding programme, there were around ninety named Benton varieties. They have wonderful names. 'Benton Apollo'. 'Benton Nigel', named for Beth Chatto's friend Nigel Scott, who died tragically young on a plant-hunting expedition with Morris in the Canary Islands. 'Benton Cordelia', 'Benton Menace' and 'Benton Farewell', the last of the lineage before Morris's eyesight failed. I spent weeks mulling over them on the Beth Chatto website before selecting

'Benton Olive'. The ground is greenery-yallery and the falls are marked with a delicate net of purple, growing darker at the throat. Like all the Benton irises, it appears to glow. Oscar Wilde could have worn it in his buttonhole.

Our garden needed a tutelary spirit of its own and that autumn I bought a statue. I rejected a bust of Capability Brown outright, dithered over a bare-breasted Bacchante, and finally plumped for Eros, armless and lovely, his hair falling about his face in snaky coils. He was a copy of a copy, a stone cast from a Roman marble, which was itself a copy of a Greek bronze attributed to the sculptor Praxiteles. At his base was the faint legend: '1378 TORSE ANTIQVE D'AMOVR. L'AMOVR GREC MUSE VATIC'. For now, he was planted ignominiously by the bins, frocked in bubble wrap, awaiting the construction of the pond and time to go to town on him.

The first stage of making the library garden was to clear away everything that wasn't needed. Matt dug the hogweed out by the wall, and between us we cut the hoheria and *Viburnum rhytidophyllum*, reducing the crown of the former and raising the skirts of the later. The tree surgeon came again, wrenching out the bamboo and stump-grinding the roots of the cherries. Ian and I spent a sweaty day digging the bed by the yew hedge clear, amassing two rubble sacks of burdock roots for the dump. The soil was dry as dust and in a full day of digging I only saw four worms. Once we'd finished, I covered the bare earth with cardboard, drenched it, and then topped it with a foot of manure, later adding two hundred worms from Wiggly Wrigglers, inserted in little groups into the earth, to bring the soil back to life. Now I'd leave it alone until winter and then plant the fruit trees, the quince and crab apple that would form the architecture of this new space.

The rest of the garden felt fully magic again, opened up and somehow mysterious after its exuberant summer season. We walked at Dunwich in the mornings, where there were horned poppies in the shingle and you could hear the bells at Walberswick windblown and mixing with the song of the larks. Home again to weed and prune, getting ready for bulb-planting time. If I stayed out too long I'd catch the rooks coming home, dozens and dozens pouring over the pond garden, wheeling on their fingertips, forming a clamouring circle just behind the coach house, the latecomers trickling in at the back.

The Bloms order arrived on 15 October, and I planted it bag by bag over the next month. This was the real beginning of my garden: *Nectaroscordum siculum*, Sicilian honey garlic, and *Fritillaria persica*, with its heavy maroon bells. *Gladiolus byzantinus*, foxtail lilies and *Narcissus poeticus*, known as pheasant eye for the fringed red cup at the centre of its white petals. I hoped it would naturalise under the mulberry, where the real pheasant liked to walk. Last year I'd mostly confined myself to tulips in pots but this year there were hundreds to go into the beds. 'Lady Van Eijk', 'Blushing Beauty', 'Doll's Minuet', 'World Friendship', 'Flaming Spring Green'. Striped 'Helmar', scarlet

'Ile de France', white 'Maureen', each placement meticulously worked out by way of endless lists.

The pond garden reeked deliciously of rotting figs. The combination here was pale creamy yellows, 'Moonlight Girl' and 'City of Vancouver', tall 'Carmargue', its white bowls finely striped in red, and a scattering of the hot-pink 'Marietta', lily-flowered like the snaky tulips on Iznik plates. The fig was knee-deep in its own crumpled yellow leaves, the colour of Ian's favourite handkerchiefs. There was a beautiful sunset after I finished, with very soft and diffuse underwater light, so that the fish swam about in pink castellations of cloud. All that was left to do with the year now was prune the roses and mulch.

My friend Philip had curated a show of Derek's work at the John Hansard Gallery in Southampton and two weeks after the last bulb went in we travelled up for the opening. There was a storm that day and all the trains at Waterloo were being cancelled, as tree after tree fell on the lines. It looked as if we wouldn't make it through, but at last a train did leave for the coast. My garden diary is almost exclusively confined to what happens within its walls, with very occasional mentions of outside occurrences, but in the next entry I described seeing Philip in his orange sweater across the length of the gallery. We embraced in the doorway and found it impossible to let go, it had been so long.

The show was called Modern Nature. It was so beautiful, so intensely moving and carefully constructed that I found I was crying as I walked around. Philip had placed Derek's films and paintings alongside work by older and younger artists, setting up a complex network of affinities in the darkened rooms. There was the Durer engraving *Knight, Death and the Devil*, and a Keith Vaughan of naked muscular boys playing an inscrutable game, looming with their ball against a troubled sky. A film on a loop kept producing poppies and burning stubble, then wartime Rome in snow. 'You were ambitious for the landscape of the English imagination', the filmmaker's voice said, and then: 'It's time to open the garden.'

Here was the twilight England Jarman had written about in *Dancing Ledge*, with its stone circles and mud-larked medieval rings, its drowned bells still audible if you listened hard enough. In another room there were the black paintings I liked so much, with their cargo of religious artefacts and elegant hand-lettering. Keith once told me Derek could write just as fluently backwards as forwards. Mirror writing, a true magician's trick. One of the paintings contained a crocodile, along with an alchemist's set of compasses.

Right at the end, in the final room overlooking the street, there was a wall of Howard Sooley's photographs. Howard was Derek's collaborator in the garden as well its documenter, his accomplice on plant-hunting trips. I stood looking at those pictures for a long time. They were small and arranged modestly in rows. There was Derek at St Bart's, his face terribly emaciated. There was a Scotch thistle on the Ness and the yellow horned poppy we also have at Dunwich. There were several photographs of photographs, taken from the Jarman family albums, so that he appeared as a boy as well as a man in his prime, standing amidst the flaming yellows of his garden.

Time was falling in upon itself. An opium poppy, palest mauve; the corkscrew spirals of anti-tank fencing, repurposed for love, not war. Mirrors, sickles, his mother's passport, the glass lenses of a lighthouse. 'I would love to see my garden through several summers,' he wrote at the end of *Modern Nature*, and by sheer force of will he did. He's been gone for so long now, but the pictures still spoke the tongue he'd taught us, defensive and salutary: the codes and counter-codes, the secret language of flowers.

VII

THE WORLD MY WILDERNESS

O
n the first day of the year I braved the ladder into the scaffolding that had gone up around the coach house just before Christmas. There was a narrow walkway all round the roof, and I worked my way along it, heart in mouth. I wanted to see the garden from above, to record some of the mysteries of its structure, which were only visible in winter. Each room was so distinctive, the lines a little off kilter, so that no corner was ever quite ninety degrees. The beech hedge was russet against the many greens, rising sinuously in its pair of arches. The box cubes in the pond garden were more mismatched than I'd realised, and every now and then I caught the sharp scent of the witch hazel, drifting upward on the breeze.

I finally got a good look at the pigeon loft, too. There was a curved wooden door halfway up the wall, decorated with rows of square-headed nails, and above it four holes set more recently in a diamond pattern, each with a little brick landing platform underneath. Mark had kept pigeons. There were photographs of them in one of his albums, fancy birds that might have been rollers. When I'd climbed the wooden ladder inside the coach house months ago I'd seen their nesting chambers, at the end of what must have been a hay loft when this was a working stable. The pigeons were long gone now. The nests had been requisitioned by jackdaws, who used the platforms as podiums for impassioned arguments over twigs. I hoped the builders wouldn't scare them away, not that they seemed scared of anything.

It's possible I was projecting. The builders started work on 4 January, tearing off the rotten garage doors at the front of the coach house, and punching a hole in the back wall. A digger trundled through on its caterpillar treads and by the end of the week the entire roof was off and covered in black plastic, and a hole had been dug for the pond, revealing a foot of black topsoil above pure sand. Somehow they'd cut the lines according to an old drawing, removing an azalea with scented yellow flowers along with everything else in the bed. I drove Ian mad worrying about compaction and exposed roots. Each time I went outside there was something new to fret about. A dozen paving stones piled against the Irish yew. A scaffolding pole jammed across the lilac, which I'd already cut back harder than I liked. I knew it was necessary, but it felt as if all the work I'd done was being reversed, to see the garden so raw and exposed.

Once again, my mood was shadowed by circumstance. A few days after the builders arrived, my father went for an angiogram at his local hospital and texted a few hours later to say he had potentially fatal narrowing of his coronary artery and needed an emergency bypass. He wasn't even allowed to go home and get a toothbrush. I'd noticed he was struggling to walk at the Chelsea Flower Show the previous autumn, its usually immutable calendar inverted by Covid. By the end of the day he was grey and panting, and I'd made him promise to call his doctor. Details, details, he'd said, deflecting my concern with a flap of his hand.

At dawn, he was taken by ambulance to St Bart's. Very rapid journey with blue light on, he texted. Please note visitors forbidden. I went to London anyway, leaving Matt on his monthly visit to plant the quince and crab apple in holes I'd dug. My father's surgery was scheduled for the next morning, a five-hour operation. In the afternoon, just as we were beginning to hope it might be over, it turned out the first patient on the list had overrun. Every time I phoned he seemed to have been moved to a new ward. He'd asked for some clothes and a document from his computer and the next morning my sister and I collected his keys from an orderly and went to Hertfordshire.

I hadn't been in his house for a long time, and never without him. We hadn't felt welcome as children, never had our own bedroom or kept any things there. The task involving his computer was unbelievably complicated. I left my sister puzzling it out, packed him a bag with enough clothes for a fortnight and went out into the garden. The light was catching in a cornus, turning it blood-red. Despite the lawyers it looked as if he was going to lose the house, and his garden with it. I was certain the stresses of the past year, combined with the gruelling decade of his wife's illness, had contributed to his situation now. Heart sick, heart sore.

When we got back to London I went alone to Bart's and dropped his bag at reception. He really was in surgery now and I spent the rest of the afternoon walking aimlessly through the City, winding my way through the smallest, oldest alleys, not quite lost but not always sure where I was; the sort of historical walk my father loved. I went into every church I passed, pausing to read the plaques or sitting quietly in a pew at the back. The first I came to was St Bartholomew the Great, the oldest surviving church in London. I lit a candle and went on by way of West Smithfield, pausing to look at my favourite memorial. It marks the site where King Richard II met with Wat Tyler and other representatives of the Great Revolt of 1381, to agree the political reforms that he later reneged on, killing Tyler in the process and later John Ball too; the events that inspired William Morris to write his time-travelling fantasy *A Dream of John Ball*. The plaque is carved with John Ball's famous words, still unfulfilled, still yearning: 'THINGS CAN NOT GO ON WELL IN ENGLAND NOR EVER WILL UNTIL EVERYTHING SHALL BE IN COMMON WHEN THERE SHALL BE NEITHER VASSAL NOR LORD AND ALL DISTINCTIONS LEVELLED.'

I went through the passageway into St Bartholomew the Less, which is set inside the hospital precincts and has a stained-glass window of a kneeling nurse in blue-striped linen and a starched white apron. Bart's is the hospital where Derek Jarman was treated and where he died, on 19 February 1994. I didn't know this at the time but a couple of weeks before his death he'd sat in this church with his friend Howard Sooley and chosen the hymns for his funeral. He recited the words to 'All

Things Bright and Beautiful' and when Howard looked over he was crying.

I didn't want my dad to die. I didn't want him to lose his garden. I said a quick, unhappy prayer and carried on, looping through Postman's Park, fragrant with Christmas box, and back by way of Bartholomew Close, where Milton hid during the frightening months after Charles II returned to London. It was comforting to think of these things, to imagine how many other people had walked these narrow streets burdened by grief or fear. If I cut up by London Wall I realised I could come out by St Giles-without-Cripplegate, where Milton was buried.

I'd been in the church, but never looked around it properly before. I found Milton's statue and then a bust of him, both clearly sightless, alongside a commemorative plaque with his dates. At its base a muscular snake was in the act of clamping its jaws around a very realistic marble apple, a simulacrum of the ruddy gold one Satan claimed 'more pleas'd my sense / Then smell of sweetest Fennel'; the apple that caused humanity so much trouble. It was amazing that these statues had survived the Blitz, considering how badly the church was damaged. There was a display about it in the corner, showing St Giles in the aftermath of the bombing, roofless and filled with rubble. Milton's statue had stood outside in those days and on 25 August 1940 he'd been blown clean off his plinth by a blast. Later I found a photograph of him lying on his back at the base of a sandbagged wall, his puritan's hat in his hand, looking in blind astonishment at the sky.

His remains were not so lucky. A hundred years after he

was buried his coffin was opened and for two days people raided the grave, wrenching out his hair and teeth to sell as relics, before this scandalous behaviour was stopped and his poor corpse reburied. The last relics were destroyed on the moonlit night of 29 December 1940, when so many incendiary bombs rained down on the church that even the cement burst into flames. It was the worst night of the Blitz in the worst affected area of London, the bombs falling so fast that an observer compared them to apples from a tree. The whole City was ablaze, a continuous sheet of flames that turned the sky a queer rose-pink and the river scarlet. By morning the roofless church was standing amidst sixty acres of smouldering rubble. Eight Wren churches went up that night and twenty million books were burned to ash, from the publishers and book dealers that had been plying their trade on Paternoster Row, in the shadow of St Paul's, since Milton was a boy. So many houses were destroyed that by 1951 only twenty-eight people were registered as living in the wasteland that had been the Cripplegate ward. In 1851, the number of residents had been 14,361.

My father was always telling us stories about wartime London, inspired by nearly every street we walked down. It was a source of fascination to him, the way the city had been so rapidly ruined and so drastically rebuilt, so that you sometimes had to squint to make out the seams of its repair, to locate the sealed air-raid shelter or vanished church. Both he and his father had spent their working lives within the not quite square mile of the City itself. After my grandfather was demobbed in

1946 he'd worked in an office on London Wall, a few minutes' walk from St Giles. When he was a young man you could criss-cross the City by way of a maze of courts and alleys, without ever having to go into the open air at all, like the mice in *The Tailor of Gloucester*. By the time he came back from the war, it had all been smashed to smithereens, so that even the shortest and most familiar journey became circuitous and strange.

In those days the bombsites were occupied by a profusion of flowering weeds, so that the ruined city resembled a garden, incongruously mantled in gold and imperial purple, the ripe smell of buddleia mixing with the sourness of brick dust and mould on the air. After the Great Fire, it was London rocket that had grown most abundantly among the ruins, but after the Blitz its place was taken by the glowing pink spires of rosebay willowherb, known as fireweed for its ability to flourish on burned ground. It spreads by means of its prodigious seeds, eighty thousand in a season, and its presence brought an abundance of elephant hawk-moths to float, pink too, in the dusk above the ruins. There were gaudy drifts of Atlas poppy, which looks as if it's made from rags of fine orange silk, as well as coltsfoot, gallant soldiers, Canadian fleabane and the pyrophile Oxford ragwort, which grows in volcanic ash in its native Sicily, and can likewise take advantage of the sites of recent fires. Most of these arrivals were wind-borne, but some seeds were brought by birds, or on people's boots, while the tomato plants that flourished in sunny places were thought to have derived from office workers' lunches. Aliens, adventives, provincials, invaders, pioneer colonists and accidental waifs:

the language used by botanists of the time was not exactly neutral.

I first heard about this sumptuous garden, rising from the ruins of the city, planted and planned by no one at all, in *The World My Wilderness* by Rose Macaulay, the novelist who also wrote a life of Milton. It was published in 1950 and set in the immediate aftermath of the war. In it Macaulay describes a London transfigured: 'a wilderness of little streets, caves and cellars, the foundations of a wrecked merchant city, grown over by green and golden fennel and ragwort, coltsfoot, purple loose-strife, rosebay willow herb, bracken, bramble and tall nettles, among which rabbits burrowed and wild cats crept and hens laid eggs.' Her vision had settled in my head, so that whenever I saw the gleaming towers of Moorgate and London Wall they always seemed a little unreal, their possession of the city no more permanent than the flowers they'd supplanted.

That day I stopped my wanderings at dusk, but it was nine before the surgeon rang to say the operation had gone off okay and my father was sleeping in the IC ward. For some reason they'd taken away his phone and laptop, and when I called the next morning he'd been moved again. Then came a series of baffling texts. High on Oramorph, he kept saying he was cold and then repeatedly sending his own name. The surgeon had told him he'd be kept in hospital for a fortnight, but four days after the operation a nurse told one of his friends he was being discharged tomorrow. It seemed unbelievably fast. I went to Bart's to fetch him, escorting him back to Suffolk in an Addison Lee cab, cheery and white-faced, with a towel to stop the seatbelt from irritating the wound where his heart had been opened and reset.

Once he was settled in bed with his laptop and a cup of tea, I scoured the house for *The World My Wilderness*. I found it at last in Ian's study: a hardback with a Barbara Jones print of weeds growing out of rubble, silhouetted against the stalwart dome of St Paul's, which had miraculously survived the raids. I hadn't read it in years. It's about a teenaged girl called Barbary, who has spent the war running wild in the south of France, on the junior fringes of the Resistance, the *maquis*. Her mother, who is very beautiful and indolent, sends her to London to live with her father, a KC, upstanding and remarried, who has no understanding or much liking for his savage daughter with her sly slaty eyes. Barbary is supposed to be studying painting at the Slade but she whiles away her days in the ruins of the City, instinctively grasping that this is a kindred place, with a kindred population, to the *maquis* where she was trained.

I read her route into the wilderness in astonishment. She enters it by way of Noble Street, one road east from where I'd walked up Aldersgate. In those days it was an alley, barely more than a footpath, that weaved through the flowering ruins, past fig trees and elders. Barbary makes her way to the shell of an unnamed church, its roof blown off, its angels gazing into the empty air. Marigolds grow in the niches and in the belfry tower there are eight broken bells alongside a bronze statue of a man with long hair and boots. But this was the statue of Milton I'd just seen! Surely this was St Giles itself? I read on, until I came to a description of a plaque of John Speed. That settled it. The entire book was located among the ruined guild-halls, the pigeon-haunted offices and law courts where my grandfather had worked, picking his bowler-hatted way between mystifying relics.

Barbary falls deeper and deeper into this ruined world, until it almost destroys her. The whole novel is engaged in an intense post-war struggle between civilisation and barbarism. Why bother, the spivs and deserters who lurk amidst the ruins seem to say. Why not steal and cheat and help yourself, why not have as much comfort and luxury as you can snatch, when death is so close and whatever the churches once signified appears to have been blasted clean away? The infesting weeds are a symbol of this malaise, but Macaulay can't quite help but find them beautiful and alluring, even though she knew better than many people what the wreckage meant. Her own flat was bombed on the final night of the Blitz and she'd had the melancholy experience of shinning up beams and girders to try and find her lost possessions, though all she recovered was a few jars of marmalade and a silver mug among the charred wreckage of her library. Every beloved book had been destroyed, including her father's 1845 edition of *Paradise Lost*, which she'd used to write her life of Milton.

It was the presence of plants that turned the bombsites from a site of tragedy into something more fertile and seething with possibility. It certainly sounded more interesting and alive than the Pret a Mangers and investment banks that infest the same area today. A garden that seeds itself, the grass turning to golden hay in August, the sodden greenery in autumn, the smell of bracken and mould and ripening figs. The milky blue stars of asters, 'the red campion, the yellow charlock, the bramble, the bindweed, the thorn-apple, the thistle and the vetch': isn't that preferable to concrete and steel? On the final pages, the wilderness is already being beaten back, the cranes and derricks on their way. The weeds are burning on pyres, but it is clear that they'll return, ceaseless and tenacious, so that for a weird unstable moment we are no longer looking at the recent past but instead forwards, into our own future, in which the city to come is itself a ghost and the wild flowers once again reign undisturbed.

I wish I could have seen those flowers. Before she's expelled from her perilous Eden, Barbary makes a scratch living painting the ruins on postcards she sells to tourists for one and six, upping the price to half a crown for Americans. I wondered what her pictures were like. Far more sloppy and impression-istic, I guessed, than the exacting images produced by the painter Eliot Hodgkin, who in the 1940s created an even more luminous and haunting record of London's wilderness than Macaulay. Hodgkin was at the time working in the Home Intelligence Division of the Ministry of Information, as well as volunteering as an air-raid warden. He could see the bombsites around St Paul's from his office window, and in 1944 he began painting them, climbing like Barbary into open basements and cellars to capture the dome of the cathedral against the shat-tered walls and improbable pink antennae of the willowherb.

In 1945 Hodgkin wrote to Kenneth Clark begging to be allowed to produce an official record of the plants that had established amidst the ruins of the City. Clark was the chair of the War Artists Advisory Committee, established in the autumn of 1939 to commission artists to produce a documentary record of Britain at war. The best known images in the collection are those made by Henry Moore and Edward Ardizzone of people sleeping in the impromptu air-raid shelters of the Underground, mouths hanging open, bodies spectral, worn to a ravelling by fear and exhaustion. But there are thousands more, by over four hundred artists, known and unknown, recording bombing raids and burning churches, first-aid posts and operating theatres; even a house in the act of collapsing. *Convalescent Nurses Making Camouflage Nets*; *An Emergent Bridge over the River Thames*; *Fire in a Paper Warehouse*; *Escape of the Zebra from the Zoo during an Air Raid Fire*: the titles attest to the uncanny and often terrible sights that were witnessed and logged for the nation.

In his letter to Clark, Hodgkin asked for work from the committee that 'will make some use of my special abilities – such as they are – and be a bit less uncongenial than the civil service.' Hesitantly, he put forward his idea. 'It has occurred to me that possibly no pictorial record has so far been made of colonization of London's bombed sites by wild flowers – willow herb, ragwort etc. If this is the case, I would like to suggest that a few of the most striking and familiar of these plants should be painted in their surroundings, before post-war rebuilding sweeps them away forever.'

The paintings he made that summer are akin to Durer's *Great Piece of Turf* in their almost miraculous recreation of the intricacies of the most disregarded wild forms. Hodgkin (an older cousin to the abstract painter Howard Hodgkin) had become increasingly fascinated by still lives of small and humble, common and overlooked objects, among them gourds, ribwort plantain, knobbly yellow quinces and kale; perhaps a testament, at least at first, to the limitations of subject matter during wartime. His paintings of the ruins, made in egg tempera on panels, convey the extreme beauty of the vegetal tapestry that emerged during those years.

Only one was accepted by the committee: *The Haberdasher's Hall, 8 May 1945*, V.E. Day. The hall, a seventeenth-century replacement for a medieval building destroyed in the Great Fire, had been bombed in an air raid five years earlier. A single tea-coloured wall survived, buttressed by broken cross walls. Hodgkin had constructed his painting so that it stood alone in the rubble, edged all round either by a soft, elaborately cross-hatched blue sky or by a tracery of golden ragwort, backlit so that the petals rang like coins – a composition that has something in common with one of his favourite medieval images, The Last Judgement from the *Hours of Mary of Burgundy*, with its bright marginal ribbon of borage and cornflowers flowing around a central icon of destruction and death.

Another painting from the series went unpurchased, though to my mind *St Paul's and St Mary Aldermanbury from St Swithin's Churchyard* is Hodgkin's masterpiece. Pink giving way to brown, orange at the edge of green, muted, autumnal colours that suggest it's possibly already September. If the Haberdasher's Hall used plants to counterpoint the agony of destruction, this painting is more like a love note to regeneration, a rapturous celebration of the beauty of flowers most commonly described as weeds, though anyone familiar with Gerard's herbal would have recognised their true nature instantly. I'd kept several of them in my old dispensary, in the archaic form of tisanes and tinctures, and even at this late juncture could still recite their botanical names. *Rumex crispus*, curly dock. *Taraxacum officinale*, dandelion. *Stellaria media*, chickweed. Medieval plants for the medieval merchants who'd once had their gardens here, returning to take up their long-accustomed stations.

There was no sense of panic in this painting, no apocalyptic *After London* terror, but instead a relaxed appreciation of *rus* reclaiming *urbe*. The city's chimney pots behind and the plants in front, each occupying their own distinctive time frame, so that what seems most transient is also closest to eternal. What was the lovely line from *Cymbeline*? 'Golden lads and girls all must, as chimneysweepers, come to dust.' *Golden lads* is said to be a nickname for dandelions and *chimneysweepers* their seed heads, though plainly not in Shakespeare's day, when the chimney brush they resemble hadn't yet been invented. Hodgkin had included both here, spelling out in the language of flowers a constant cycle of decay, regeneration and return, in which we all play a part.

My father was not a good patient. He was up at all hours, pounding at his laptop, never sleeping, impossible to persuade back to bed. The builders had lined their hole with breeze blocks and moved onto other things. Ian's library was proceeding, but my garden was in stasis. Walls kept coming down and going back up, and with them showers of mortar that I went out at first light to fish from the umbilicus of ferns. Hellebore, daphne, snowdrop, but everything felt out of sorts and in a muddle. I took refuge in reading, losing myself in Ian's hefty book of hand-coloured maps of London's bombsites. The site around St Giles was almost entirely purple. I consulted the key: damaged beyond repair.

The bombsites were a developer's dream, of course. Many became car parks and then office blocks and luxury housing, literally paving the way for the glass and steel London of today. But the wild garden that formed in the ruins also presented its own possibilities, suggesting to a diversity of people the appeal of a greener, more fertile metropolis. Many of the City's ruined churches were turned into public gardens, among them St John Zachary, the graveyard full of sunflowers in *The World My Wilderness*, now an appealing sunken garden with glossy magnolias and *Hydrangea petiolaris* trained up its walls, St Dunstan's in the East, and St Mary's Aldermanbury, the Wren church that Hodgkin included in his painting of St Paul's, which Barbary races past as she tries to escape the police. These days it's a shady enclosure on the corner of Love Lane, planted with knots of box and shields of yew beneath two copper beeches.

A flurry of new parks in the East End were likewise established on bombsites or land reclaimed during the post-war clearance of so-called slum housing, much of it badly damaged in the Blitz, including Shoreditch Park, Haggerston Park, Mile

End Park and Weavers Fields. The housing estates that replaced the slums were often designed with integrated gardens too, some of them beautifully thought out; part of the post-war settlement that for a time reversed the trend of the enclosures, returning a little of the common wealth of Britain to its poorest citizens. Housing, education, the welfare state: in those years there was a brief vision of the nation as a collective garden, in which everyone could partake of the fruit.

Perhaps the most ambitious project in this utopian burst of building is the one that rose from the ruins around St Giles, the site of Barbary's wilderness and Milton's grave. The Barbican is the largest art centre in Europe and an estate housing over four thousand people; a marvel of brutalist architecture that is also liberally appointed with gardens for residents and visitors alike, its concrete balconies gaily decorated with pennants of scarlet and pink geraniums. It embodies the concept of public luxury, its enticements ranging from cinemas, theatres, an art gallery and a library to a tropical conservatory with fifteen hundred species of rare and endangered plants, as well as ghost carp, grass carp and terrapins. The centrepiece is a formal lake, its green waters populated by carp and patrolled by herons, built over the bombed warehouses of Fore Street and Redcross Street fire station, which itself burst into flames during the worst night of the Blitz.

I was only really working in the garden at weekends now, rocketing out at dawn and spending all day clearing leaves for leaf mould, uncovering snowdrops and the first aconites in their yellow surplices and grass-green ruffs. I picked the snowdrops daily, entranced by the solidity of their whiteness, as dense and shining as porcelain on their green scapes. The bees liked them too. I found one feasting, bright orange pollen overflowing from its baskets. I lifted this year's crop of lamium, replanting violets as I went, 'Who Do You Think You Are Kidding Mr Hitler' looping absurdly through my head. Failing light, scraping out shepherd's purse and old robinia leaves, breathing the damp odours of late January.

February was intermittent, notes scratched as I went back and forth to London. Blood and bone the box, feed the magnolia, plant a 'Félicité-Perpetué' rose on the library wall, a Valentine for Ian one day late. Storm Eunice arrived as I was dividing snowdrops, bringing down two branches of the magnolia. Later I heard a crack and looked up through the window to see a pine falling in the park behind. I'd been thinking so much about war and then on 24 February Russia invaded Ukraine, and the news was full of bombed cities and displaced people, walking to the Polish border laden with pushchairs and animal carriers. Overnight even the smallest Suffolk villages had Ukrainian flags, snapping blue and yellow in the breeze.

War is the opposite of a garden, the antithesis of a garden, its furthest extremity in terms of human nature and human endeavour. It is possible that a garden will emerge from a bombsite, but it is certain that a bomb will destroy a garden. In the first horrified week after the invasion, I watched several times a short film called *The Last Gardener of Aleppo*. It was made by Channel 4 in May 2016, at the height of the Syrian Civil War, and told the story of Abu Waad, who ran the last surviving garden centre in rebel-held Aleppo. The city was at the time being continuously bombed, both by the Syrian regime and the Russians, who were now using what they'd learned in Aleppo to attack the cities of Ukraine.

The film was made during a lull in hostilities. It opened with Abu Waad like Adam naming his plants: hazelnut, loquat, pear tree. He showed the interviewer a tree that had been hit by a barrel bomb, said that it would survive, said with total certainty: 'I own the world. We ordinary people own the world.' He grew his plants in tins, cajoling the most luscious roses out of red sand. Some of his wares, he said, were bought by his neighbours to decorate the roundabouts of Aleppo, which has been inhabited for longer than any other city on earth. This planting was a statement of defiance, an act of construction amid the wasteland of ruined buildings and open craters in which they lived. We saw one, planted with pots of parsley and what were surely snapdragons. Then the camera turned to Abu Waad's thirteen-year-old son, Ibrahaim, who was hunched and terrified, unlike his jolly father, as he cut roses for a customer visiting the local hospital.

Six weeks after this sequence was filmed a bomb landed near the garden centre, instantly killing Abu Waad. Ibrahaim was interviewed again, though this time he could barely speak, was somewhere far beyond where words apply. Then the film cut back to Abu Waad in his garden in the spring, smoking, drinking tea, pruning a rose. 'Flowers help the world,' he said, 'and there is no greater beauty than flowers. Those who see flowers enjoy the beauty of the world created by God. The essence of the world is a flower.'

Perhaps it is enough in wartime for a garden simply to exist; a signed affidavit that there are other ways of spending a life, that generosity, gentleness, cultivation also count, despite their perpetual vulnerability. A war garden provides a kind of sustenance even when it is not given over to cabbages and carrots, to allotment plots and Dig for Victory. And sometimes a garden can be a literal refuge, too. Its gates can be opened, it can turn from a private to a shared sanctuary. That spring I found myself thinking often of a garden I'd once come across in Italy: a lush, aristocratic garden composed of descending terraces and fountains, enclosed box hedges and cypress trees, so green and orderly on its hillside that it looked almost alien above the harsh, bleached landscape of the Val d'Orcia, a region of Tuscany set roughly midway between Florence and Rome. In the 1940s, a tin box was buried in this garden, containing a diary that recorded – day by day and sometimes hour by hour – the hostilities and terrors that convulsed this region during the Second World War; a testament to the many different purposes a garden can serve.

When Iris Origo first saw La Foce, on a stormy day in October 1923, she was twenty, a clever, wealthy, not particularly happy Anglo-American girl raised in Italy, fatherless, and newly engaged to a penniless Italian aristocrat a decade her senior. The house was, in her estimation, not beautiful (she had been raised, after all, in the echoing immensities of the Villa Medici in Fiesole, on the hills above Florence, where the rooms were still papered with the yellow Chinese silk installed in the eighteenth century by Horace Walpole's sister-in-law). The estate was huge: 3,500 intimidatingly desolate and dusty acres, so badly neglected that it was almost derelict. The land itself was suffering from erosion and deforestation; 'pale and inhuman', it reminded Iris of the mountains of the moon. She and Antonio spent three days riding around this wild place with the *fattore*, land-manager, visiting the outlying farms, all twenty-five of which were in an advanced state of disrepair. There were few roads, no electricity. Much of the land was fallow. Most seriously of all, there was very little water.

Neither of the Origos had any experience with farming. What they did have was energy, ideas, drive – and, in Iris's case, a deep well of money. She brought to the marriage an inheritance of $5,000 a year, as well as regular gifts from her American grandmother. In the mid-1930s, she received a second unexpected inheritance from a distant cousin, which took her from the ranks of the wealthy to the seriously rich. Part of the bond between the Origos was that both were powerfully motivated by a desire to be useful, to make more of a difference to the world than the idle, indulgent circles in which they'd been raised. 'There will be a great deal for Antonio to do', she wrote eagerly to a friend after seeing La Foce for the first time, 'and I shall "visit the poor", play the piano, run the school, I hope write – not a bad life.'

She sounds like Dorothea in *Middlemarch*. The paternalism of her fantasies was perfectly matched both by the existing system of Italian agriculture and by the country's new government. It was a year almost to the day since the Fascists had marched on Rome and seized power, the menacing mobs of Blackshirts in the driving rain. Socialists, communists, anti-fascists, journalists had been beaten or sent into exile. Some had been killed. But the Origos' dream chimed with Mussolini's new programme of rural restoration, the *bonifica agraria*; the rebuilding of what he regarded as the traditional Italian countryside. In those early years Antonio entertained Fascist dignitaries at La Foce, dressed like them in a black shirt. Both he and Iris encountered Mussolini at the opera, at balls and parties; were sometimes beckoned over for a private conversation. Their acceptance – queasy or not, expedient or not – of Fascism was common among their class (Churchill was as late as 1927 among the charmed), but it persisted long after many of their closest friends set themselves against it, at great personal risk.

The agricultural system in Tuscany was then the medieval *mezzadria*, akin to sharecropping in America, which had been briefly reformed after the First World War before being re-instated by Mussolini. This is how Iris explains it in her 1970 memoir *Images and Shadows*:

. . . a profit-sharing contract by which the landowner built the farmhouses, kept them in repair, and supplied the capital for the purchase of half the live-stock, seed, fertilisers, machinery etc., while the tenant contributed, with the members of their family, the labour. When the crops were harvested, owner and tenant (the date I am writing about is

1924) shared the profit in equal shares. In bad years, however, it was the landowner who bore the losses and lent the tenant what was needed to buy his share of seed, cattle and fertilisers, the tenant paying back his loan when a better year came.

She thought it worked well, on balance, as long as both landlord and tenant were conscientious; 'the bad aspect of the system, from the first, was that an idle or self-indulgent landowner, who did not repair and stock his farms, crippled his peasants, too.'

I'd read the word *mezzadri* before, and not long ago. I found the passage in *Thin Paths*, Julia Blackburn's memoir about life in a remote village in northern Italy. A neighbour is telling Julia about her childhood. She explains that she and her family were *mezzadri*, and when Julia doesn't understand she crosses one index finger over the other, to make it half. 'Almost everyone in the village', Julia writes, 'was *mezzadri* or half-people, which meant that they owned nothing for themselves. They belonged to a *padrone* who was their master and they had to give him half of everything they produced, down to the last kilo of olives or chestnuts, the last egg or cabbage.' This neighbour was five when her father explained the system to her, saying: 'We are nothing and we own nothing.' Seeing that he was upset by his words, she tried to argue with him, pointing out that she wasn't half a person, and he lost his temper with her, though he was usually very calm and quiet.

There's no doubt that the Origos were the good kind of

landowner. It is as if the many benefits they brought to the valley – the roads and lakes, the schools, the clinic and nursery, the modern farming equipment, the rebuilt houses and greener, infinitely more productive land – formed a pile so vast that it somehow obstructed their ability to see that no matter how well you treat the people that you own, you are still the person with absolute authority over their lives, and that this will remain a source of pain however helpful or generous you are. Iris had grown up inside it, after all: a world of masters and servants, people with money and people in need, as her Irish grand-mother once observed, of buttons and teeth.

The American branch of her family, the Cuttings, derived their fortune from railroads and sugar beet, which her grand-father and great-uncle were the first in America to refine. The Cuttings were named in the Four Hundred, the list of those wealthy and well-bred enough to belong to New York's exclu-sive Gilded Age society. Her grandfather was a founder member of its most prestigious institutions, among them the New York Botanical Garden and the New York Public Library, the Metropolitan Museum and the Opera.

If the Cuttings belonged to the rarefied world of *The Age of Innocence* (Edith Wharton was a family friend), then the other side of Iris's family might have come trotting out of a novel by Molly Keane, that remorseless chronicler of the Anglo-Irish Ascendancy, the adults on hunters and the children on ponies, their dogs milling at their heels. Iris's grandfather was Lord Desart, and as a small child she'd spent her summers at the family estate, Desart Court in Kilkenny, which she describes in

Images and Shadows as 'our earthly Paradise'. She thought she could still find her way around it blindfolded – the study with its gentleman's library, the servants' hall, the bluebell wood, the mysterious shrubbery where the children had their camp, the sun-ripened peaches on the kitchen garden wall – though it was long gone by the time she came to write her memoir, burned to ashes by the IRA the same year that Mussolini came to power. In the chapter about her grandfather, she insists again and again upon his decency, his reticence and good manners, somehow incapable of understanding that no matter what a good, kind man he was, his paradise remained an occupation on someone else's land, provisioned by someone else's labour. Philanthropy was a tradition with both sets of grandparents, a way of continually attempting to ameliorate a kind of wealth that at some buried level must have been understood as indefensible.

It was her Cutting grandmother who'd made the garden at La Foce possible, providing the newlyweds with a pipe that carried water to the house from a spring in a beech wood six miles away. Almost as soon as the taps began to flow Iris and her architect friend Cecil Pinsent started constructing their garden on the hill, making it one green enclosed room after another, as funds allowed. The two had known each other for a long time. Pinsent was famous, at least among the English residents of Florence, for his elegant interpretations of the Italian Renaissance garden, inflected with a whimsical touch of the English Arts and Crafts. He'd restored the elaborate gardens at Villa Medici for Iris's mother Sybil. It was the oldest surviving example of an Italian Renaissance garden, and it's not impossible that Milton himself visited it during his long stay in Florence. He spent the summer of 1638 in the city, a trip which is supposed to have inspired his vision of Eden as a wilderness, far closer to an Italian garden of the time than anything he could have seen in England.

Iris's garden at La Foce began simply enough, with a lawn surrounded by box-edged flower beds, embellished at the centre with a fountain decorated with stone dolphins, where her small son Gianni liked to play with his wind-up frog. By the early 1930s, there were walled gardens and flower gardens descending the hill like the decks of a liner; a working kitchen garden, pergolas with roses and wisteria, and wilder woodland gardens with flowering quince and species roses, underplanted with bulbs, among them daffodils and bluebells, the latter of which disappointingly failed to thrive in the clay. Pinsent laid out the design, the walks and terraces and balustrades, making a living structure of box and travertine, and Iris embroidered it with the flowers she ordered in bulk from English nurseries: tulips, irises, dahlias, sweet peas and above all scented red roses. It was a place to be private, to think and walk and talk with her friends. In a letter to her grandmother she said she was 'entirely absorbed by the loveliness of my flowers.'

It would be wrong to think that she was just a gardener in those years. She was making her school on the estate, and conducting an affair with Colin Mackenzie, the friend to whom she had confided her Dorothea-ish fantasies of helping the poor, which ended abruptly when Antonio discovered their letters. She spent enjoyable hours playing with her son, raised otherwise, as she had been, by a nanny, and travelled almost perpetually, drifting round Europe in a haze of parties and suppers and opening nights. The Greek Islands, the Grand Hotel in Venice, tea in Rome with Lytton Strachey and Lord Berners, skiing in the Maurienne. It was on this latter trip that Gianni contracted the illness that killed him fifty-eight days later, meningitis.

This vast, unfathomable loss – 'I can't pretend that it seems anything but an utterly empty world' – ushered in a new era of seriousness. She wrote two biographies in the 1930s, had a second affair, almost broke away to England for good. She'd finally started to notice what was happening in the world. Friends were fighting against Franco in Spain. Even she couldn't ignore the iniquities of Mussolini's war in Abyssinia. At the end of August 1939, she went with Antonio to Lucerne for a series of concerts, conducted by Toscanini and played by his orchestra of Jewish refugees. The last was *Götterdämmerung*. In the hotel that night they heard on the radio that Russia had ratified the Non-Aggression Pact with Germany. Foreigners began to flood out of the city. She drove across the frontier with Antonio, watching as the pole of the barrier swung down conclusively behind them. The next day Germany invaded Poland, and on 3 September 1939, war was declared.

She was eight months pregnant when the Battle of Britain began. She listened to it on a radio hidden under her pillow in a nursing home in Rome, watching the moon over the domes as she imagined London and what might be happening to the many people she loved there. The baby was born on 1 August. That autumn she got a job with the Italian Red Cross, locating prisoners of war. For the next two years she spent her weeks in Rome and what weekends she could spare from work at La Foce with her new daughter, Benedetta. This pattern didn't change until 30 January 1943, the day she wrote the first entry in the diary that was later published as *War in Val d'Orcia*: the diary she kept hidden among the books in her children's nursery before moving it with her jewellery and propaganda leaflets to a tin box buried in the garden. It is an extraordinary document: a cool, dispassionate, sometimes wry and in its later portions wholly terrifying account of invasion, occupation, liberation and all the grievous changes they wrought.

The Origos might have been model landowners, but the distinction between them and the peasants they were trying to help was absolute. With the war those borders started to dissolve. The garden in particular became a more open space, a refuge for many more people than Iris and her intimates. *War in Val d'Orcia* begins with the arrival of refugee children at La Foce, 'seven very small, sleepy bundles . . . like small bewildered owls.' They came from Genoa, where their homes had been bombed and they had spent the last two months living with their family in the grim confines of an underground tunnel. All were thin and malnourished; several plainly traumatised. On 10 January, six more girls arrived from Turin. Eventually there would be twenty-three refugee children at La Foce, billeted in the nursery school, playing in the rose garden, splashing in the fountain.

All spring the Allies were bombing Italian cities. On 3 May the Italians requisitioned Castelluccio, the castle a mile away from La Foce that the Origos had bought in the 1930s, for fifty British prisoners of war, most of them from Yorkshire. Iris kept away from them, knowing that she would be able to provide more aid if she stayed discreet. She was pregnant again, though she barely mentions it in the diary, too caught up in the news and mounting daily tasks. On 9 June a second daughter was born, again in Rome, this time between air raids. It was unnerving, she wrote, listening to them from her hospital bed, where she'd given birth while listening to the screams of a young airman who was having his leg amputated.

Food was scarce in the city but the flower stalls were awash with roses, irises, Madonna lilies. There were political arrests. This time she recorded them. On 10 July the Allies landed in Sicily, and the next day they had a christening party for Donata in the garden. Rome was bombed. On 25 July, the radio reported that Mussolini had resigned, though the news took a day to reach them. 'A weight has been lifted,' she wrote, 'a door opened; but where does it lead?' The first German officers appeared at La Foce on 4 August, 'tramping up the garden path'. More appeared as she was playing blind man's buff with the children in the garden. It was her birthday and they proposed a formal toast.

On 17 August the first bombs fell in the Val d'Orcia and on 3 September the Allies arrived on the mainland. The Italian armistice was on 8 September, a lovely still morning, and the next day the Allies landed at Salerno. She'd hoped there would be an Allied landing near Rome and was beginning to realise that the German front would be driven north and that things could get very bad indeed. She walked up to Castelluccio and gave the English prisoners of war the news that they were now technically free, suggesting they escape immediately. By 10 September the Germans had occupied Chiusi, ten miles away. The phone lines were cut. No post, no buses, no papers, just the radio to connect them to the world. She and Antonio buried supplies of wheat, potatoes, cheese and wine, along with petrol. They took the wheels off their cars and hid them. News came that the Germans had occupied Rome. The chances of a successful escape were diminishing by the hour and the Origos decided it was wiser to hide the English prisoners of war before the Germans arrived, scattering them between the more remote farms on the estate.

The woods around the house were full of deserting Italian soldiers and many more escaped Allied prisoners. Iris opened the garden gate to them, assisting everyone who came to ask for food and directions. In this, the Origos behaved very differently to other Italian landowners, almost all of whom supported the Fascists. All of Italy north of Rome was now under German martial law. On 12 September they heard the bells of St Paul's ringing on the BBC for what she wrote ironically was 'victory' in Italy. On 16 September there was an announcement: anyone who continues to feed or shelter British prisoners of war after twenty-four hours will be subject to German martial law. On

21 September, a reward was added: L1,800 for capture of a British prisoner of war or information on their whereabouts. On 26 September Florence, Pisa, Verona, Bologna were bombed by the Allies, and a new order awarded the death penalty for anyone sheltering or assisting members of enemy forces. All the time reports kept coming through of trains packed with terrified Italian boys, on their way to labour camps in Germany.

The hidden English prisoners of war were becoming a serious problem. Many people in the neighbourhood knew of their presence, and on 28 September two German officers came to the house to tell Antonio that he had to report their presence. Asked what would happen to them, he was told they would most probably be sent to camps in Germany. That night he and Iris sat up in bed discussing what to do, with the German officers still in the house. They hurried out at dawn and warned the prisoners' leader, Sergeant Knight, who formulated a plan for escape that wouldn't implicate the Origos. Iris gave him her last slab of chocolate and a map.

On 4 October they walled up their linen, blankets, silver in thirty-two labelled boxes. Anti-fascists were being arrested in Montepulciano. On 3 November the Jews were deported from the Rome ghetto. Those who hid in attics and cellars were known as *sepolti vivi*, the buried alive. It snowed for the first time on 8 November. Everything would be harder now. She spent her days in a hundred jobs, entertaining the children, making clothes and nappies out of old counterpanes and curtain-linings and slippers out of strips of carpet.

The Villa Medici, the house in which Iris grew up, so lovely that Princess Mary had borrowed it for her honeymoon, was being requisitioned. She found it impossible to regard its loss as important, though she got permission to remove its most valuable furniture and to seal up the drawing room with its yellow wallpaper. 'As I go from one familiar room to another – all now full of German soldiers – I have a strong presentiment that this is the end of something: of this house, of a whole way of living. It will never be the same again.' People were starving in Florence, and everywhere she went she heard that Jews and anti-fascists were being deported. At home, the young men on the estate were running away rather than report for duty to the German army. She found out that her gardener's son Adino had snuck back to say goodbye to his father, spending the day hiding in the garden. When she told Gigi that this was 'very unwise', he broke down and wept. The Germans had begun arresting the fathers of recruits who refused to report for duty.

On 27 November she decided to list the day's fugitives, giving a record of the kind of people continually appearing at the garden door: Italian soldiers; British prisoners of war; desperate, destitute groups of evacuees, one suffering so badly from pernicious anaemia that Iris sent her to hospital in Siena. They wanted boots or a map, bandages, warm clothes, shoes, a place to sleep, books, sometimes a job until the Germans left and they could go home. Two soldiers, escaping from a concentration camp; a family of Jews, for whom they found a hiding place in a remote convent. People were being tortured in

Florence. Every day she had to calculate her difficult equations: how many people to help, how much risk to take, who was being endangered by each seeming act of aid. 'Life', she wrote, 'is returning to the medieval pattern: as the outside world is more and more cut off, we must learn, not only to produce our own food and spin and weave our own wool – but to provide teaching for the children, nursing for the sick, and shelter for the passer-by.' By the end of the year a thousand people were living in the chestnut woods around La Foce, depending for their survival on the Origos and the *mezzadri* farmers.

In January a thousand German paratroopers arrived in Chianciano, requisitioning oil, wine, sheep. There were plans to billet one hundred soldiers at La Foce and the castle. The Origos packed up furniture and mattresses and sent them to the remotest farms by ox cart. They discovered they were under Nazi surveillance as suspect persons for giving funds to partisans and encouraging peasants not to join the military. The Allies were thirty miles from Rome. They hid a radio in the nursery to listen to the BBC, now banned. Increasingly, there were partisans in the woods, the Italian equivalent of Barbary's *maquis* in the south of France, sleeping at the remoter farms or in secret camps on Monte Amiata.

A vignette from that winter captures the extremities of Iris's life at the time: 'Find on my breakfast tray a note from the nurse: "Have got a man here with a bullet through his shoulder, who killed a Fascist last night. What should I do with him?"' She was saving lives, but she was still a *marchesa* who had her breakfast brought up on a tray. A partisan boy died of influenza and as they said the service in the little chapel where her son Gianni was buried she realised the partisans were standing on the hill, looking like brigands in a novel. After dark they came down and laid flowers on the boy's grave. She supported the partisans, but they often angered her with their looting and risky attacks on the Germans, which inevitably led to reprisals on farmers and villagers.

On 12 April she found two unfamiliar English prisoners of war lounging on deckchairs in the garden, more confident than the usual terrified visitors. They warned her to be careful, saying that many people on their route had told them to take refuge with the Origos. That month the Germans tried again to requisition Castelluccio and La Foce, for three hundred men and eight officers, as well as stabling for eight hundred horses. A steady stream of partisans flowed into the woods. Iris was denounced in a Fascist paper. Leaflets were dropped from German planes describing what would happen to people who helped rebels.

By 3 June the bombing and machine-gunning on the roads was constant, from dawn to dusk. On 5 June the Allies entered Rome, and the Germans took over Castelluccio and the schools. The refugee children were moved into the house. The garage and courtyard were full of German Red Cross lorries, their exhausted drivers asking for coffee and food. 9 June was Donata's first birthday. They had a party in the garden for the children, running three-legged races while the planes swooped overhead. The next night Iris buried her papers in the tin trunk in the garden.

Strange scenes as the front draws near. They are rehearsing *Sleeping Beauty* in the garden when armed soldiers burst in, asking the children to sing 'O Tannenbaum'. They are invited to join the partisans, who ask Antonio to be mayor of the district after liberation. The courtyard is still full of German soldiers, who wash naked in the laundry while the partisans slip unnoticed into the cellar to get wine. At midnight Iris and Antonio walk through the garden. 'It is a strange sight – the great bulk of cars, filled with sleeping men, concealed behind every arch and under every tree or thicket, camouflaged with young cypress which they have ruthlessly cut down . . . We wander about among it all, with a curious sense of detachment, feeling like ghosts of the past, who have no business here.'

Now they are rehearsing *Snow White*. They no longer need to ask where the front is. They are the front. A battalion of German paratroopers arrive, pick locks, steal their possessions, including mattresses and Iris's sunglasses. Three farm girls are raped. The soldiers tell her England is being bombed by a new weapon, that London is already destroyed. She thinks continually about where it will be safest to hide the refugee children, when the Allies and their bombs arrive. A trench in the woods? The cellar? They have a cow for milk for the babies, but it escapes. There are machine guns behind the parapet that Cecil Pinsent built, and the roads are mined. A shell bursts in the garden.

She writes her diary while boiling milk for the children in the kitchen. Everyone is hiding in the cellar, including sixty people from the outlying farms who have spent the night being shelled in the woods. There is a lull in the bombing and the Germans order them to leave. They go at once, a terrified crowd, with a pram full of nappies and a basket of food. Iris's own suitcase contains the following items: underwear for her and Antonio, shoes, soap, eau de Cologne, face powder, a clock, and a photo of Gianni. They stagger, not quite running, down a road they know has been mined. It is very hot and there are uncovered corpses. The planes come back and they lie in the corn. The group divides. Some leave for Chianciano and the Origos and sixty others continue to Montepulciano. Twenty-eight are children, carrying their winter coats, and four are babies. After walking for four hours, they rest beneath the rampart of the town. Suddenly people come running to meet them, among them the familiar faces of partisans and refugees they'd helped.

It takes a week for the Allies to reach Montepulciano. The Germans shoot a partisan and his body hangs for days from a lamppost, sickening everyone. They hear that the garden of La Foce has been bombed. Then the Allies arrive and Iris finds herself at what she describes as 'a party in a dream', drinking wine and eating biscuits with British officers. It turns out they have friends in common. The next day, her cousin Ulick appears, with whom she used to play in the bluebell woods of Desart Court.

She finally got back to La Foce on 1 July, in a staff car. The garden was shocking. It was full of shell holes and trenches, and was covered in a foul litter of broken objects, among them torn-up mattresses and what she recognised as her own private letters and photographs. The lemon trees had been pulled from their pots and left to die. The gardener Gigi, who loved flowers and took such risks for his son, was found dead in a ditch, killed by a shell. Inside, the house stank of rotten meat and human shit. The lavatories were all blocked, and there was broken glass and broken furniture and everywhere the torn pages of her books. Fifteen farmhouses had been destroyed. There was an infinitude of work ahead, and yet the war had finally passed away from their door. In a letter written that summer, she wrote: 'How strange it is to be able to *look forward* again – to see the horizon widening, instead of narrowing . . . Most of the garden is beans and cauliflower, but there is still jasmine on the wall and there are still fireflies in the corn.'

My father had a second week of convalescence with my sister and then went home. We tried to find him a carer and a cleaner but he refused all help. He lived not far from the Beth Chatto nursery and at the end of the month I met him there for lunch. The builders were nowhere near finished but I needed to start planting. Back in the autumn I'd said we'd open the garden in June for the National Garden Scheme, also known as the Yellow Book. The NGS organises the opening of around three thousand private gardens across the country to raise money for charities, many of them to do with nursing and caring, and one of which supports veteran soldiers to retrain as gardeners.

We were what was known as a resurrection. Back in Mark's day the garden had always been in the Yellow Book and I wanted to do him proud. I had a long list of plants to find and after a sausage roll each my father and I went to the nursery. The first task was to locate 'Benton Olive'. Then I piled a trolley with an abundance of shade-lovers: rusty foxglove and Japanese anemone, *Anemone nemorosa* and *Pulsatilla vulgaris*, *Iris pallida* and sweet woodruff. I added what I described in my diary as *a couple of interesting comfreys*, not knowing then the havoc they'd wreak, *Tellima grandiflora*, a plant I love for its strange pale yellow cup-like flowers, and a polemonium, pale lilac. My father bought me a melianthus as a thank you for the past weeks of care and we finished with tea and flapjacks and a discussion of the latest dismal twists in his case.

Planting, my entry for the next day began. It took from nine in the morning till four in the afternoon. The manure and cardboard I'd laid in the autumn had worked its magic. The soil was rich and heavy, unrecognisable from last year's thin brown sugar. I planted a scattering of dark red martagon lilies too, and the next day the builders took down the scaffolding and finished the last row of bricks on the now mended arch. I planned to cover it in clematis; a purple macropetala and a yellow tangutica, with their delicate downturned bells. Already the garden was a sea of bird song. When I drove to Halesworth that afternoon there was a light green glow over the fields and the trees and the air was warm and hazy blue. Spring was coming, in grotesque contrast to the news from Ukraine.

The papers were full of images exactly like those in Iris's diaries. Bombed cities, exhausted people crossing borders, pushing prams piled with clothes and bread, their children and sometimes even dogs in rucksacks on their backs. In the haste and confusion of the evacuation from La Foce, Iris had forgotten to fetch her dogs from the kennels. It was one of the unbearable things that came back to her as they rested in the fields on the way to Montepulciano. When they did get home they found their pointer Alba dead in the fountain. But when she went into the stinking house she saw a black shadow under a sofa and a moment later her poodle Gambolino crept out, thin, nervy, deaf, but very much alive.

Though the story of *War in Val d'Orcia* is one of unalloyed heroism, it has a sad coda. The book ends with Iris paying tribute to the 'patience and endurance, the industry and resourcefulness' of the *mezzadri*, the peasant farmers who had risked so much to help the partisans and prisoners of war. 'Resigned and laborious, they . . . turn back from the fresh graves and the wreckage of their homes to their accustomed daily toil. It is they who will bring the land to life again.' She thought that amidst the horrors of the war the old barriers of class had finally been broken down, that a new, 'local' solidarity had formed. But by the time she came to write *Images and Shadows*, two decades later, this relationship had soured. It turned out the eternal suffering and stoicism of the peasants that Iris so admired was not, in their estimation, eternal at all. Like the Diggers, they wanted change, and particularly change in the *mezzadria* system. As soon as the war was over, almost all of the farmers at La Foce converted to what she describes somewhat sourly as 'the new doctrine which promised that, if they obeyed orders, they would swiftly come to own the land on which they worked.'

Communism, that burning word Goodwyn Barmby had carried from Paris to England so many years before, had reached the Val d'Orcia. The farmers fought for an increase in their percentage of the wheat, from 50 to 53 to 57 per cent. It was resisted by the landowners, who claimed it wouldn't leave them enough profit margin to improve the estates, never mind that many of them never had. There were strikes, and tenants who were evicted refused to leave the houses that their families had occupied for centuries. Iris wondered if at root the problem was one of communication, that the barriers of class were preventing people from speaking honestly and openly to each other. She always preferred to see people as individuals, which is part of her charm, and part of what is so frustrating about reading her work. She could not understand that there might be something systemically, even fundamentally, wrong with one person owning land on which many other people depended for their lives, no matter how kind or well-intentioned they might be.

For a time it seemed as if the post-war Italian government would pursue a policy of land restitution, breaking up the great estates and allowing the farmers who worked there to purchase their own farms, a dream expressed succinctly as *La terra ai contadini*. Instead, what happened is what happened all over Europe. The people left their farms and moved to the cities, and agriculture entered a new, depopulated phase. The next stage in La Foce's story was relayed three decades on by Iris's daughter Benedetta, in a lushly illustrated book about the garden and its history. She explained that the 'acrimonious' years in which the *mezzadri* were so angry ended when the Origo sisters sold off a third of the estate, which was bought by the state and run as a co-operative, first by the original *mezzadri*, and then, when they 'sadly ... went bankrupt through mismanagement and lack of experience', by a second co-operative of Sardinian shepherds. A national park was set up by five of the Val d'Orcia communes, and a new prosperity settled on the region, fed this time not by agriculture but by tourism. Many of the farmhouses on the estate, where partisans and prisoners had been hidden by farmers at such great risk to all their lives, were converted into luxury villas, while La Foce itself could be rented for weddings and christenings and film shoots.

I was sitting on my green sofa one night, not long after my father had left, when I saw it on the television. The long ochre house, the green box hedges, running like mascles through a shield: it was unmistakably La Foce. The garden was the setting for a party, in a drama about the hyper-rich, the day before one of the family got married. People were moving between the borders with glasses of champagne, climbing the terraces to drip poison in each other's ears; people who had come by helicopter and private jet, people with infinite reserves of money. I was watching with my mouth hanging open. There was the youngest son, tossing his drink over the parapet where the German soldiers had set their guns. The war had been won all right, and it was not as Abu Waad had said. The ordinary people who own the world; when do they get the keys?

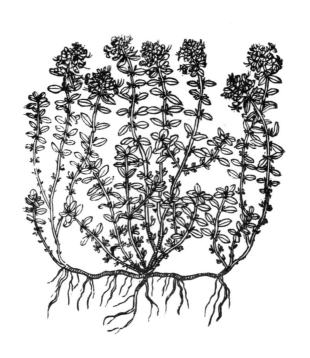

VIII

THE EXPELLING ANGEL

The tulips had come, floating above the soft blue clumps of forget-me-not and purple knapweed. The magnolia had put on its full regalia. May, the light draining away by 8.31, a smoky blue iris opening beneath the fig. According to repeated entries in my diary, I could hardly tear myself away each evening. Ian's garden was coming on in fits and starts. For months there'd been a neatly bricked hole where the pond would be. Then one day two men came and covered it in fibreglass, which had to harden before we could fill it. A few days later I came home from London to find the circle around it had been covered in hardcore, as a prelude to paving. Idiotically, it hadn't occurred to me to check the levels. I'd been imagining a slightly sunken garden but when they began to lay the slabs next morning it became clear that it was set like a stage three inches above the beds. Too late, oh well. All that planning.

It was fine in the end, of course. The stones looked beautiful. Rees and the two Jamies manoeuvred Eros into position in the baking sun. I spent the evening digging out buckets of mortar, rocks, cans, nails, string from beds compacted after months of scaffolding and diggers. We were opening the garden for the NGS on 11 June. A month to go, and this part of the garden was still only half-planted. I kept going back and forth to the local nursery, picking up irises and nicotiana of many kinds, plus a few trays of white cosmos I couldn't resist. I put in the 'Benton Olive' and my father's melianthus, as well as plants I'd lifted from other beds: violets and hellebores, clumps of yellow Welsh poppies. Even unfinished it was magical to be in, especially alone. It was so quiet and enclosed, constantly visited by birds, the scent of the wisteria ghosting through. When I went back for a last look at dusk, there was a toad by the door to Ian's library, blinking in the half-light, his throat throbbing.

On 13 May the electrician arrived to wire the library lights and I begged him to put a plug on the pump so we could fill the pond. A few hours later Jamie came to get me, to ceremoniously turn on the tap. Brimful, the pond changed the space in ways I hadn't reckoned on, exerting the mysterious gravity of a new element. The next morning I got up very early, to wander around on my own. The medlar was covered in splayed white flowers, which look so much like Tudor roses. I clipped a few shrubs very lightly, trying to give everything its own defined shape. Later that day we put in the pond plants: the marsh marigold that John Clare called *horse blob*, yellow flag and water plantain. I sank a water lily on a milk crate I'd found in the potting shed, loving the coolness of the water on my arms, and planted *Iris pallida* in front of the Christmas box, for sweetness summer and winter alike. There was a mining bee burrowing in the border, and later I saw a mouse in the greenhouse, darting past as I watered the pelargoniums.

Everything was ahead of itself. March had been like May and now May was like midsummer. Roses, delphiniums, lupins, peonies were all in flower, and the irises were almost over. After rain the rose border was full of glowing heads of brilliant apricot, pink and deep wine-red. God knows what would be left by June. There was nothing I could do about it, no way of slowing that mysterious timepiece. I planted foxtail rosemary by the library steps, an idea I'd stolen from my friend Simon at Worcester College, where it lolls across a flight of steps onto the quad, sticking up in electric-blue plumes. My diary was full of increasingly deranged lists of things to do. Every task I ticked off seemed to spawn two more, a never-ending frenzy of planting and tidying. I was coming to the end of two years of work and I didn't want a single element of Mark's garden to feel neglected or unthought about, even if my version was wilder than he might have liked. *Use pots to distract, deadhead roses, hope for the best,* I wrote sternly to myself.

The centrepiece of the NGS garden is the tea. We'd been planning to serve it in the old stable yard, which the builders still hadn't vacated. At last, with a week to go, they packed away their things. I swept it clear and without really meaning to spent a blissful afternoon lugging in hoses and wheelbarrows and big pots of roses, bumping them up the steps on Ian's sack trolley. Ian helped me move a table over, and I set out pots of aeoniums and pelargoniums, finally liberated from the greenhouse. By the end of the day, I had the working yard I'd been dreaming about for weeks, if not years. I was so excited to be able to get into the compost bins again I practically jumped in them.

Three days to go, two days to go. There were road signs to put up, on stakes we sharpened with an axe. Ian made fifteen cakes and forty-eight fairy cakes and insisted on icing them all himself. We borrowed two trestle tables, a big box of china and two tea urns from the village hall, and fetched a bag of change from the bank. I mowed the lawn, deadheaded everything in sight, got stung by a bee as my reward. The first opium poppy opened, a true scarlet with ragged petals that looked as if it had been attacked with prinking shears. That evening I saw the summer's first bat. The builders finally finished the last of the snagging. The last thing they did was nail two horseshoes above the yard doors. A flurry of white Dutch irises had opened in the library garden, along with white martagon lilies I was sure I hadn't planted and had no record of ever buying. Mark's revenants perhaps, making an unscheduled appearance.

And then it was the day itself, warm and still, with the deep blue sky I'd been praying for. The garden looked immaculate, which is to say wild and abundant with colour and scent. I set out every chair we owned in convivial groupings, the bench under the shade of the magnolia, striped deckchairs by the hazel. Both my parents came over. My friends Rebecca and Sam had matching sun hats. They were doing the first tea shift, followed by Margaret and Lorraine. At 10.50 I was sure that no one would come, and at 10.55 a chatty little queue was building by the gate. I had my roll of paper tickets, my box of change. The garden was open again.

It was probably the best day of my life: just for the feeling of looking in and seeing the garden so full of people, talking to each other, making themselves at home. It was like a party all day. People came in for a minute and stayed for hours. Dozens had visited before, or were old friends of Mark. They stopped me as I went back and forth between teas and gate, to describe what it had looked like in his day. In the pond garden, a man in a panama hat told me Mark had worked at Sissinghurst and that Vita herself had given him the white fig in the corner. He said Mark had been involved in the restoration of Giverny too, painstakingly identifying the flowers in Monet's paintings so they could be reinstated, though when I later contacted the house they denied that any English gardener had been involved. By the end of the day I felt as if I'd done right by Mark; as well as any amateur could, that is. I'd never be able to attain the heady perfection of his borders, but I'd done my best.

And that would have been the end of it, the punctuation mark, except that it stopped raining. Even back in March I'd been fretting over the lack of rain, the dryness of the ground. *Still no rain*, I wrote in April, and *so dry this year* in May. East Anglia is the driest region of the country. It gets around half the annual rainfall and has been categorised by the Environment Agency for over a decade as 'seriously water stressed'. By mid-May I was regularly logging yellow leaves and flopping plants.

June was alarming, and my diary for 20 July was apocalyptic. There had been two days of record temperatures. It reached 40.3 in Lincolnshire, while in Suffolk the high was 36. Airports had closed because the runways were buckling, and a school went up in flames after the sun was concentrated through a chandelier. There were more fire callouts, I read, than at any time since the Blitz.

That afternoon I walked around the garden, making an unhappy audit. Tree peony flagging, hydrangea leaves scorched, the lawn toast, not that I cared about that. The heat hit me in the face as I opened the door. I'd never experienced anything like it in England. I left out saucers of water for the wasps and mice. Mulberries were ripening by the hour, though the tree looked stunned and sick, its tips dying back, its leaves limp. A scattering of hot rain at nine. There were froglets in the little meadow under the plum, and a grasshopper jumped into my hand, the kick of its legs instantly familiar from childhood, though I hadn't seen one in years.

The paper that day also contained a story about mass seabird deaths and the unwelcome news that permission had been granted for the nuclear power station Sizewell C, despite a recommendation from the planning inspectorate to reject it. Sizewell is where we swam, next to Minsmere bird reserve. There's been a nuclear power station there for decades but planning for the new one had been rejected because it had emerged that EDF, the power company involved, had no plan in place for the water supply. Sizewell C would require 2 million litres of potable water a day to cool the reactors and the irradiated fuel, and as much as 3.5 million a day during the construction phase. There wasn't enough local groundwater to supply it, and a plan to run an eighteen-mile pipeline from the river Waveney was ruled out when it transpired the water company's abstraction licence was about to be downgraded by 60 per cent in order to keep the river flowing.

Under normal circumstances the planning inspectorate's rejection would have been the end of the proposal, but the secretary of state overrode its verdict. 'The secretary of state disagrees with the examining authority's conclusions on this matter', the government's decision letter said, 'and considers that the uncertainty over the permanent water supply strategy is not a permanent barrier to granting consent to the proposed development.' Six weeks later the secretary of state, Kwasi Kwarteng, would be appointed chancellor of the exchequer, and thirty-eight days after that he would be sacked, having brought the pound to a record low against the dollar, increased borrowing rates on mortgages and caused economic chaos that would take years to mend. It was a denial of reality that had gone on far too long, and that summer its grim consequences were everywhere I looked.

By August we were officially in drought. It was predicted to last until October or even into the new year. My dominant feeling during those rainless weeks was a kind of listless horror. The plants were dying and I could choose to water, since water was still coming out of the tap, as long as I chose to disregard the consequences, the rivers that were drying up by the day. The source of the Thames moved five miles downstream. The profligate way we used water began to feel actually insane, the kind of thing I'd look back at in a decade or two and find it hard to believe was once so normalised and ordinary. Swimming at Walberswick I saw clouds of smoke roll up from the horizon near Dunwich. Somebody's fields on fire. You could see the burn sites from the train, black earth and stubble. There was no grass for the cattle to graze and the farmers were already using up their winter feed. I found a dead toad in the garden and later a dead baby jackdaw, pushed from its nest.

On 17 August a hosepipe ban was announced. I'd stopped using the hose a few weeks earlier. I couldn't bear to water my garden, to make my piece of land more important than any other, and I also couldn't bear to see the plants I'd cherished die. The ground had actually baked. It was rock-hard, so that any water just ran off, particularly where I'd cleared between plants. The days were cooler now and overcast, the sky thinly veiled in grey, but it would take weeks of rain to soak the soil. Even Vita's fig was struggling, its fruit floury and dry. The mulberry harvest, on the other hand, was spectacular. We kept tracking bloody footprints through the hall. I made a running tally of which plants were surviving and which could perhaps no longer cope without a level of summer watering I couldn't bring myself to do. Woodruff, astrantia, mahonia, lady's mantle, anything newly planted. The roses had blackspot from the stress and when the dahlias came into flower they were shrunken and deformed. Nearly all the trees had patches of dieback now, while the Lavalle hawthorn with honey fungus was almost entirely skeletal.

Maybe I should start again: replant entirely for drought, or mulch a foot deep instead of an inch, so the sandy soil would hold more moisture. Every article I read, and I read dozens, suggested water butts, but a water butt contains 200 litres. Forty watering cans. You'd empty it in a few days, and then what, in the baking months of no rain? I'd long since given up on baths, my former passion. I investigated rainwater storage tanks. In London no one was talking about the drought, though there were dying trees in all the squares, their leaves burned as if by fire.

One of the most horrifying aspects of those weeks was seeing the garden become another manifestation of selfishness, a private luxury at a shared cost, rather than a place that ran counter to the world's more toxic drives, a refuge from its priorities, where other forms of life are given more regard. In my years on environmental protests, gardening had seemed the best, least damaging way to spend a life. Even my decision to become a herbalist had stemmed from a conviction that growing plants was among the most ethically permissible things to do. But a garden is many things at once, as I'd begun to see: selfish and selfless, open and enclosed.

It was during this frightening time that I began to read a sequence of garden poems by Milton's friend and colleague Andrew Marvell, the man who was probably responsible for preventing his execution. Marvell wrote them when he was still a young man, in his late twenties or very early thirties, during the unsettling final years of the English Civil War. He was living then in rural seclusion at Nun Appleton House in

Yorkshire, working as a tutor to Mary Fairfax, the daughter of the retired soldier General Lord Fairfax, who had led the Roundhead troops until he was replaced by Oliver Cromwell.

The 'Mower' poems give voice to a violent ambivalence about gardens and their place in human life. I'd read them many times, but they struck a very different chord during those hot unhappy weeks. The one I liked best was 'The Mower Against Gardens', a satirical account of the garden as a corruption of nature by luxurious and seductive man. Roses 'taint' themselves with 'strange perfumes'; the white tulip learns to 'interline its cheek', like a girl putting on blusher for the first time. Man grafts wild plants on tame, creating forbidden mixtures, producing strange sterile hybrids. It's a bizarre vision of the garden as sexual, deviant, dangerous, double, foreign, its plants enchanted and drugged by a wicked enchanter.

> He first enclosed within the gardens square
> A dead and standing pool of air,
> And a more luscious earth for them did knead,
> Which stupefied them while it fed.

Marvell's tongue is firmly in his cheek here. The speaker of the four 'Mower' poems is a censorious rustic, perennially suspicious of artifice, a failed lover who elsewhere rails against his wounding rejection by the lovely Juliana (in 'Damon the Mower', he is carrying on in this vein when he accidentally cuts himself with his own scythe – 'the mower mown', a line that had often made me laugh while mowing the lawn). At the same time, those lines did encapsulate how I was starting to feel about the greed of gardens. *A more luscious earth*: peat extracted from bogs and packed in plastic sacks, shipped in vans from Amazon delivery hubs, the drivers watched by a computer that set them schedules in which there is no time to rest or eat. The invisible cost of every single thing you buy, no matter how benign your intentions for it.

The corrupted garden of the 'Mower' poems is by no means the only type of garden in Marvell's poetry, of course. He returns to them continually, as sites of refuge and rapture, as a right-thinking alternative to war as well as a place of degenerate desires. The pinnacle of all these works is 'The Garden', that complex, layered argument about retreat and contemplation. I'm not sure there is anything in literature that better enacts the spell of being in a garden than the middle stanza, in which the plants are weirdly more active than the speaker and time itself is ensorcelled, slowing line by line until it stops dead, stunned:

What wond'rous life in this I lead!
Ripe apples drop about my head;
The luscious clusters of the vine
Upon my mouth do crush their wine;
The nectarine and curious peach
Into my hands themselves do reach;
Stumbling on melons as I pass,
Ensnar'd with flow'rs, I fall on grass.

In this poem, the garden becomes by turn a place of seclusion, delight and repose, a doorway into the secret universe of the imagination, which humans, perhaps alone of all the animals, occupy simultaneously to the material realm. Man is, as Thomas Browne had written a few years earlier, 'that great and true *Amphibium*', living in two worlds at once: the visible and the invisible. Marvell here essays a version of the same thought. His garden is a place for dreaming of things not yet created; 'Far other worlds'. It inevitably provokes a comparison with the happy garden-state of Eden, but it is also explicitly postlapsarian, after the Fall, in that death is present and time is passing. In the final stanza, so often carved on sundials, the garden itself becomes a clock, the flowers ticking off the minutes and the hours, attended by bees.

The garden as a clock: what a beautiful image. Garden time is not like the ordinary time in which we live. It's different to a watch or the glowing numbers on an iPhone lockscreen. It moves in unpredictable ways, sometimes stopping altogether and proceeding always cyclically, in a long unwinding spiral of rot and fertility. To pay attention to the garden as a clock means entering a different relationship with time: as circular, not linear, as well as the acknowledgement that one of its recurring stations is death. *Et in arcadia ego*, I'd written in my diary months before. Now, as the plants died around me in such apocalyptic ways, it was beginning to dawn on me that I'd got the relationship between death and the garden all wrong. I'd somehow internalised the idea that a good garden is a deathless garden, in a state of continuous perfection, even as I'd rejected the notion of it as a sealed sanctuary, a refuge from the outside world, with its plagues and wars. But neither of those things were possible. The garden was always engaged in a dance with death. It couldn't possibly replicate Eden: that fecund paradise where the apple fruits and flowers at the same time; with, as Milton puts it, 'gay enameld colours mixt'. All this time I'd been resisting its lesson, chasing the high of perfection, feeling a failure when things browned or died back. It was as if my job was to maintain the visual illusion, as if the garden couldn't possibly look good unless I'd succeeded in excising any evidence of death.

What a strange thing to do. This was the more sinister legacy of Eden: the fantasy of perpetual abundance. I was beginning to see what a poisoned fruit it truly was. So many of our most ecologically deleterious behaviours are to do with refusing impermanence and decay, insisting on summer all the time. Permanent growth, constant fertility, perpetual yield, instant pleasure, maximum profit, outsource the labour, keep evidence of pollution out of sight. The secretary of state's refusal to accept that there was no water for the power station was the epitomisation of this mindset, and the drought its consequence. It was all catching up with us now. To accept the presence of death in the garden is not to accept the forced march of climate change. It is to refuse an illusion of perpetual productivity, without rest or repair: an illusion purchased at a heavy, soon unpayable cost, inaugurating a summer without end, the fields burning, the trees like stones.

The first proper rain came at the end of August. It was then that I remembered Mark had written a book called *The Dry Garden*. I wished I'd found it earlier. It was as if he was speaking calmly from the past into just this anticipated moment. It was published in 1994. Even then Suffolk was a dry county, and especially problematic if you gardened on sand. Mark talked about water as a precious resource that had to be conserved. His strategy was twofold: add organic matter and mulch the soil so it holds moisture, and use plants that are specialised for drought conditions. I knew both those things, but to see it explained in such thoughtful depth was very reassuring. I pored over his suggested lists. Rock rose, phlomis, artemisia, rosemary, helianthum, even *Rosa rugosa*, which grows naturally in sand. Like Iris Origo I might have to let go of delphiniums, and perhaps even some roses, but there were plants that would survive the future, the hotter, drier summers of anthropogenic climate change.

As the days shortened and the shadows grew longer, the garden became very alive again, just as it had each previous autumn. It felt luxuriant and ripe, weirdly alert. The cyclamen came back, followed by the colchicum, the now familiar ticking of this particular clock. I collected hollyhock seeds to sow in the gravel. The spent flowerheads resembled purses, each one containing a stack of tiny dark coins. Plants I'd thought had given up the ghost revived, growing enormous with the rain and cooler days. I mulched the remainder of the library beds with cardboard and manure. The little frog came back. There was good news from my father too. It was starting to look as if there was a way for him to keep his house. I expected him to be ecstatic but he surprised me by saying that it was too big for him and that he was ready to move on. What about your garden, I asked, and he said that he thought at seventy-five he had it in him to make one more.

Impressed by his resolve, I decided the time had come to replant the pond garden. It had been the worst affected in the heatwave, and had looked barren in August every year since we moved in. The beds had long since been overtaken by the more thuggish plants, especially cardoons and echinops, with its spiky blue heads, and the soil badly needed organic matter. The surface mulching I'd done the previous winter hadn't solved the problem, but I'd been scared of doing anything more substantial because there were so many plants and especially bulbs I wanted to keep. It really needed to be lifted and remade, a daunting task, though I was encouraged to find in an essay by Mark that he'd acknowledged its necessity two full decades earlier.

For weeks I fiddled around with planting plans, trying to design a border that would keep producing interest all season round, using both the plants that were already there and a few new additions, designed to weather hot, dry summers. This time, I wasn't after the illusion of perfection. What I wanted was succession: a community of plants that fitted together in a tapestry, so that each new plant to emerge naturally took up the station of the last. In this I was inspired both by Mark and by Christopher Lloyd, the exuberant creator of Great Dixter, who with his head gardener Fergus Garrett had pioneered the art of succession in a border.

The formal garden at Great Dixter in East Sussex is one of the most aesthetically dazzling things I've ever seen, a masterpiece of abundance. It was planted with maximum ornamentation in mind, a brain-melting mix of colour and form, which plays with shifting size and perspective through time in startling ways. After Lloyd died in 2006, Fergus let the garden get even looser, phasing out pesticides and artificial fertilisers altogether. It seemed to support an enormous amount of life, and not long before the pandemic the Dixter team commissioned a comprehensive biodiversity audit. The estate is only six acres, but it's comprised of many different elements, including woodlands, pastures, ponds and meadows. All of these areas were found to have unusually high levels of biodiversity. To the astonishment of the participating ecologists, who had been dubious of the merits of examining a garden in this way, the richest site by far was the formal ornamental garden. 40 per cent of the UK's bee species were logged within a year, including some that are very rare, like the long-horned bee and the white-bellied mining bee, alongside a multitude of birds, butterflies, moths, nationally rare spiders and invertebrates.

It was thrilling to discover that a place designed for beauty and abundance might also prove so deeply hospitable. The dense borders mimicked natural plant succession, providing a constant supply of food, while the botanical diversity supported many different species. Some areas were neglected and rarely touched, while others were regularly disturbed, just like the London bombsites that became such a rich habitat after the Blitz. Old trees and rotting wood weren't automatically removed, providing nesting sites and habitat for beetles and boring insects, so vital a part of the web of life that invisibly sustains our own. Good news for my Lavalle hawthorn with honey fungus, which could stay in situ and provide habitat for the imperilled organisms that feed off dead wood, rather than being chopped down and burned.

It was the antithesis of the selfish garden. As Fergus said, 'once regarded as part of the problem, gardens can now be seen as part of the solution'. All it required was a shift in focus. Instead of seeing the dead and dying plants in a garden as ugly elements that spoiled the picture, or weeds and insects as interlopers that didn't belong, they could be understood as components of a living tapestry, which hummed with energy even when some parts were too small to see. At the same time, the human element didn't have to bow out altogether. Unlike in the more austere models of rewilding, the gardener was integral. It was their aesthetic vision, their work, their decisions as to what to encourage and what to cull that made it so inviting to other types of life.

This way of looking at the garden also shifts its status as a closed space. It could be private, intimate, deeply individual, and wide open at the same time. Each garden run along these wilder, richer lines participates in a great network: a quilt made by many hands, spread out across cities and villages, encompassing private gardens, parks, allotments, balconies and verges, every square different, each one sustaining and supporting life. It was the first utopia I'd encountered, in all my searching, in which self-expression and the pursuit of beauty truly served the commons, instead of sabotaging it.

This is not to say that we don't need large-scale land redistribution, or to improve garden access, to make it an integral part of every city, every housing project. We do. What the Dixter story made me imagine was closer to William Morris's vision of a culture that has prioritised orchards of apricot trees over gleaming office blocks and luxury towers, that has planted roses in Endell Street and made Kensington a forest. Parks instead of new airports, allotments over motorways, a grand reinvestment in our public resources, an understanding that the garden, like the library and the hospital, is what makes all of our lives possible. We need gardens and the life they support established everywhere if we are to survive, and they must extend beyond the private realm, to form part of a cherished common wealth, while retaining their intimate and wayward qualities, where individual creativity can flourish.

I set to work in the pond garden in a much more cheerful frame of mind. It was October now, cooler, the ground damp after recent rain. Each morning I assembled my kit: a wheelbarrow full of garden compost and a big stack of twenty litre pots, a fork, hand fork, trowel and secateurs. It was a mammoth job, by far the most intensive I'd carried out so far. I dug up and potted every single plant I wanted to keep, keeping the bigger ones in their own soil and shaking the smaller ones free so they could be packed with others in compost. It was fascinating to see all the roots: the knapweed like rubbery jellyfish, the lady's mantle in great slabs that could be snapped apart in my hands, the Canterbury bells very delicate hairs, like foxgloves. A robin watched as I worked, hopeful of worms.

The combination of the yellow fig leaves and the purple monks-hood was so vivid it made my head spin. By the end of the week I had a little nursery under the acacia of well over a hundred plants I'd salvaged and could put back once the bed was cleared. There were hundreds of bulbs of different sizes too, including bluebells and those beautiful blue scilla. I stored them in pots too, crossing my fingers that they'd survive a few weeks.

The Claire Austin order arrived on 19 October. I'd chosen lots of plants recommended by Mark. There was *Phlomis tuberosa*, with its hooded lilac flowers, and *Phlomis russeliana*, which is similar but yellow. There was *Eryngium* 'Pen Blue' and *Achillea* 'Pink Grapefruit', lots of silvery artemisia and a *Geranium sanguineum* 'Cedric Morris'. I bought seeds of the deep blue *Nigella* 'Miss Jekyll' for Derek and *Ferula communis* and *Verbascum olympicum*, which tower over the long border at Dixter. There were even some Chaynee oysters, better known as asters, for William Morris. They all came in nine centimetre pots, ideal for establishing in autumn. I parked them in the yard, tucking the tallest in at the back.

On 8 November Matt levered out the most deeply rooted of the thugs, reducing to single clumps the echinops and cynara that had spread right through the bed. Once it was clear of everything bar aconites and roses he dug it over, reporting that the soil was a mix of clay and sand and much better than he'd hoped. I started the next morning at 8.30, barrowing on fifteen loads of manure, followed by four of leaf mould. I turned it lightly, to encourage self-seeding. Then I set about the planting, putting the plan I'd drawn and coloured onto one of the box squares, so I could consult it as I worked. The plants I'd lifted hadn't died, but had instead made fantastic new roots. Plants, I had to keep reminding myself, are so much more resilient than I seemed to think. The whole job took until 3.30, gulping down a single cup of coffee and skipping lunch altogether. *Tulips, bulbs, self-seeders tomorrow*, I wrote in my diary. Ian was away and I was back out at seven, in filthy joggers and plimsolls. I slotted in a nice muddle of salvaged Canterbury bells and primroses, and then planted all the bulbs I'd lifted and an abundance of new ones too: Marietta tulips, *Gladiolus byzantinus*, *Allium sphaerocephalon*, to run bright rivulets of colour through the bed.

When I'd finally tamped in the last plant, I sat on the edge of the pond and looked at it for a long time. Almost every plant that had been there was back, in better conditions than before. There were days of rain forecast ahead. The dahlias had blackened and fallen magnolia leaves were piling in the borders. I had the feeling of exhausted satisfaction that I only knew otherwise from finishing a book. There would always be things to do, things to change, but I had made my garden. It was full of bats and toads, bees, voles and mice, jackdaws and swifts, even the heron that had stolen the fish. It was mine and it wasn't mine. I dug in it, and so did the mole, and both of us had an effect. I still wanted to tidy it up, to manage my worries by way of exerting order wherever I could. I'd probably always be a bit like that. But I'd finally understood that a little untidiness was far more fertile than perfect borders. I could see that the skin of dead leaves and sticks under the hazel had its own loveliness, protecting the soil from drying out, nourishing microbial activity, feeding the new green snouts of the day lilies. Death generating life, evidence of our fallen state. Maybe that was better than paradise.

I wanted to linger in the garden for days, luxuriating in what I'd laid in for spring, but as soon as I'd finished the last bulbs I had to go to Italy for work. I started in Milan, went to Turin and Venice and then got the late morning train to Rome. It was Milton's youthful journey in high-speed, each leg taking hours instead of weeks. I hadn't been in Italy since the pandemic. When I arrived at Rome station, I went out into the Piazza dei Cinquecento. There was a garden I'd wanted to see for a long time and if I hurried, I could make it before it closed.

It was warm for November. I didn't really need a coat. The sun was catching in the umbrella pines, turning the needles from green to gold. I crossed at the corner, plunging through a torrent of traffic, and walked up to the Palazzo Massimo, the museum of Roman history and artefacts, which was transformed into a military hospital during the Second World War. The room I wanted was on the third floor. There was a boy in there alone, maybe fourteen, in a plaid shirt, his backpack set beside him on the black leather bench. I could feel he didn't want me there, that I was an intruder in his contemplation.

He was standing at the centre of a garden, a painted garden, made two thousand years ago for Livia, the wife of the Emperor Augustus. In 36 BC these frescoes had decorated the underground *triclinium*, the dining room of the Villa Livia at Prima Porta, just outside Rome, giving dinner guests the illusion that they were eating outdoors, surrounded by real plants and trees. Similar rooms were found at Pompeii, decorated with roses and feverfew and figs so ripe they were splitting on the branch. In the nineteenth century the Villa Livia was excavated and this marvellous dining room discovered. In 1944 the house was damaged by a bomb, and after the war the decision was taken to remove the frescoes to the museum. The painted garden was dismantled, cleaned and installed here, lining all four walls of the gallery.

I turned from tree to tree, bewildered by the totality of the experience, its startlingly lifelike quality. Lemons and pomegranates, a blackbird alighting in an olive. There were birds everywhere, birds I knew and birds I didn't, among them goldfinches and thrushes, flying over bushes of myrtle and laurel. There were roses and poppies in the grass, as well as humbler plants like daisies, hawkweed and periwinkle, the first plant I'd learned to identify in the convent garden. Everything was flowering and fruiting together, of course. Perpetual abundance: the dangerous firework of paradise.

Here and there the paint was peeling, so that between the grass plat, the palisade and the wilder region beyond there were vacancies, where time had worn the garden away. It had endured so long, surviving wars and bombs, new religions, new regimes. A partridge settled at the base of a holm oak. It was a garden out of time, a garden against time, continually occupying the present tense, insinuating itself into the future by way of each new visitor. I understood how it worked now, how gardens self-seed. Each one I'd lingered in – La Foce, Prospect Cottage, the Eden of Milton and Cranach – had informed the garden that I'd planted, making it more associative and rich. The ones I'd never seen and only read about had nourished me just as firmly as time spent in any real garden.

There's no point looking for Eden on a map. It's a dream that is carried in the heart: a fertile garden, time and space enough for all of us. Each incomplete attempt to establish it – St George's Hill, Benton End, the London of *News from Nowhere* – is like a seed that travels on the wind; like the seed of rosebay willowherb that floated in its tens of thousands across the bombsites of London, rooting itself in what seemed like the most inhospitable terrain. I'm not sure it can be stopped, however grave the forecast. Sometimes it hibernates, preserved in a bank of words. There is a line I love, from the fourteenth-century *English Psalter* of Richard Rolle. *This boke is cald garthen closed, wel enseled, paradyse ful of all appils.* But this book is a garden opened and spilling over. The common paradise, that heretical dream. Take it outside and shake the seed.

NOTES

I: A DOOR IN THE WALL

2   '*strengthen and comfort the heart very much*': John Gerard, *The Herball, or Generall Historie of Plantes* (A. Islip, J. Morton and R. Whitakers, 1633 [1597]), p. 741.

2   '*The Wild Marigold is like unto . . .*': ibid., p. 741.

6   '*fast by hanging in a golden Chain . . .*': John Milton, ed. Barbara K. Lewalski, *Paradise Lost* (Blackwell, 2007), p. 66.

7   '*I'll never forget my excitement . . .*': Mark Rumary, in Alvilde Lees-Milne and Rosemary Verey, eds., *The Englishman's Garden* (Penguin, 1985), p.120.

7   '*with a small orchard of decrepit apples . . .*': Lanning Roper, 'An Ingenious Cottage Garden', *Country Life* (11 April 1974), p. 872.

9   '*Over the years there has been . . .*': Mark Rumary, in Alvilde Lees-Milne and Rosemary Verey, eds., *The Englishman's Garden*, p.121.

12   '*All paradises, all utopias . . .*': Toni Morrison, *PBS News Hour* (9 March 1998).

13   '*The Middle Ages have formed . . .*': Derek Jarman, *Modern Nature* (Century, 1991), p. 207.

14   '*chained*': ibid., p. 77.

15   '*some gardens look like retreats . . .*': Ian Hamilton Finlay, in Robert Gillanders, *Little Sparta* (Scottish National Portrait Gallery, 1998).

II: PARADISE

23   '*dangling apricocks . . .*': William Shakespeare, *Richard II*, Act 3, Scene 4.

24   '*an unweeded garden . . .*': William Shakespeare, *Hamlet*, Act 1, Scene 2.

24   '*One feather, and the house . . .*': Virginia Woolf, *To the Lighthouse* (Penguin, 2000 [1927]), p. 151.

29   '*from those often seen . . .*': Mark Rumary, in Alvilde Lees-Milne and Rosemary Verey, eds., *The Englishman's Garden*, p. 121.

30   '*unto whom we owe . . .*': Thomas Browne, *Hydriotaphia and The Garden of Cyrus* (Macmillan, 1929), p. 83.

35   '*that low door in the wall*': Evelyn Waugh, *Brideshead Revisited* (Penguin, 1962), p. 163.

40   '*pretty box*': Thomas Ellwood, *The History of the Life of Thomas Ellwood* (J. Phillips, 1791), p. 212.

40   '*milked*': Joe Moshenska, *Making Darkness Light* (Basic Books, 2021), p. 3.

40   '*the pestilence then growing hot . . .*': Thomas Ellwood, *The History of the Life of Thomas Ellwood*, p. 212.

40   '*his dark materials*': John Milton, *Paradise Lost*, p. 63.

40   '*darkness visible*': ibid., p. 13.

41   '*the happy seat . . .*': ibid., p. 46.

41   '*With head, hands, wings, or feet . . .*': ibid., p. 64.

42   '*Paradise,* Adams *abode*': ibid., p. 90.

42   '*pure now purer*': ibid., p. 96.

42   '*So clomb this first grand Thief . . .*': ibid., p. 97.

42   '*Beds and curious Knots*': ibid., p. 98.

43   '*sweet Gardning labour*': ibid., p. 101.

43   '*to reform /Yon flourie Arbors*': ibid., p. 109.

43   '*What we by day . . .*': ibid., p. 222.

44   '*guiltless of fire*': ibid., p. 227.

44   '*casual discourse*': ibid., p. 223.

44   '*Two Paradises 'twere in one . . .*': Andrew Marvell, 'The Garden', *The Poems of Andrew Marvell* (Pearson, 2003), p. 158.

NOTES

45 *'Half spi'd, so thick the Roses . . .'*: John Milton, *Paradise Lost*, p. 228.

52 *'murderous Traytors . . .'*: John Evelyn, *The Diary of John Evelyn* (Oxford University Press, 1959) p. 412.

52 *'moving bitterly . . .'*: Rose Macaulay, *John Milton* (Gerald Duckworth & Co., 1957), p. 118.

52 *'fall'n on evil dayes . . .'*: John Milton, *Paradise Lost*, p. 176.

54 *'A Wilderness of sweets . . .'*: ibid., p. 130.

55 *'a hellish and dismall Cloud'*: John Evelyn, *Fumifugium, or The Inconvenience of Aer and Smoak of London Dissipated* (W. Godbid, 1661), p. 5.

55 *'Parsimonious Emmet'*: John Milton, *Paradise Lost*, p. 189.

56 *'The World was all before them'*: ibid., p. 332.

III: A LANDSCAPE WITHOUT PEOPLE

65 *'quiet, unassuming man . . .'*: Tony Venison, 'Mark Rumary: Someone You Knew', *The East Anglian Garden Group, Newsletter 101*, December 2010.

69 *'strange mutations'*: W. G. Sebald, *Rings of Saturn* (The Harvill Press, 1998), p. 265.

70 *'Estates of this kind . . .'*: ibid., pp. 261–262.

72 *'It was – it really was . . .'*: Diana Athill, *Alive, Alive Oh!* (Granta, 2015), p. 19.

72 *'The Cedar Walk had been planned . . .'*: ibid., pp. 23–24.

73 *'One man, one great man . . .'*: Horace Walpole, *On Modern Gardening* (Pallas Athene, 2004), p. 27.

77 *'I also give unto . . .'*: *Oxford English Dictionary*.

79 *'An open country . . .'*: Horace Walpole, *On Modern Gardening*, p. 59.

79 *'All gardening is landscape-painting'*: Joseph Spence, *Anecdotes, Observations and Characters of Books and Men, Collected from the Conversation of Mr. Pope and Other Eminent Persons* (W. H. Carpenter, 1820), p. 144.

81 *'No sooner was this simple . . .'*: ibid., pp. 42–43.

84   '*and that sweet man John Clare*': Theodore Roethke, *Selected Poems* (Library of America, 2005), p. 111.

85   '*Gardeners Parish Clerks and fiddlers*': John Clare, ed. Eric Robinson and David Powell, *John Clare By Himself* (Carcanet, 2002), p. 34.

85   '*tho one of the weakest . . .*': ibid., p. 4.

85   '*was for no other improvement . . .*': ibid., p. 5.

86   '*and what with reading . . .*': ibid., p. 11.

87   '*the continued sameness . . .*': ibid., p. 13.

87   '*the varied colors . . .*': ibid., p. 38.

89   '*I have been so long a lodger . . .*': ibid., p. 162.

91   '*the great landowners . . .*': Karl Marx, *Kapital*, Vol. 2 (Dent, 1934), p. 803.

91   '*the land-grabbers . . .*': George Orwell, *The Complete Works of George Orwell*, Vol. XVI (Secker & Warburg, 1986), p. 336.

91   '*the culmination . . .*': E. P. Thompson, *The Making of the English Working Class* (Pelican, 1968), p. 239.

92   '*until I got out of my knowledge . . .*': John Clare, *John Clare By Himself*, p. 40.

93   '*desolate*': John Clare, ed. Margaret Grainger, *The Natural History Prose Writings of John Clare* (Clarendon Press, 1983), p. 207.

93   '*All my favourite places . . .*': John Clare, *John Clare By Himself*, p. 41.

94   '*flye round in feeble rings . . .*': John Clare, ed. Eric Robison et al., *Poems of the Middle Period, 1822–1837*, Vol. 5 (Clarendon Press, 2003), p. 107.

94   '*at my back door . . .*': John Clare, ed. Mark Storey, *The Letters of John Clare* (Oxford University Press, 1985), p. 380.

95   '*thick set . . .*': John Clare, ed. Margaret Grainger, *The Natural History Prose Writings of John Clare*, p. 193.

95   '*litter in yellow heaps . . .*': ibid., p. 193.

95   '*the hard nicknamy system . . .*': ibid., p. 195.

96   '*peeping*': ibid., p. 216.

96   '*joalers . . . prison*': ibid., p. 234.

97   '*Ah what a paradise begins . . .*': John Clare, *John Clare By Himself*, p. 36.

98   *'hard fare & bad weather'*: John Clare, ed. Mark Storey, *The Letters of John Clare*, p. 600.

98   *'sultry'*: John Clare, *Poems of the Middle Period*, Vol. 5, p. 6.

98   'foot-foundered and broken down': John Clare, *John Clare By Himself*, p. 263.

99   *'bad Place'*: John Clare, ed. Mark Storey, *The Letters of John Clare*, p. 654.

99   *'purgatoriall hell'*: ibid., p. 657.

99   *'Bastile'*: ibid., p. 654.

99   *'Where flowers are . . .'*: Jonathan Bate, *John Clare: A Biography* (Picador, 2004), p. 516.

99   *'Incessantly . . .'*: John Clare, ed. Mark Storey, *The Letters of John Clare*, p. 656.

99   *'I very much want . . .'*: ibid, p. 664.

99   *'You never tell me my dear . . .'*: ibid., p. 665.

99   *'Plumbs Pears & Apple Trees . . .'*: John Clare, ed. Margaret Grainger, *The Natural History Prose Writings of John Clare*, p. 346.

100   *'I am still fond of Flowers'*: John Clare, ed. Mark Storey, *The Letters of John Clare*, p. 677.

IV: THE SOVRAN PLANTER

All information about the Middleton family otherwise unattributed derives from the extensive collection of their family and estate papers held by Suffolk Archives.

112   *'barly sugar candied lemon . . .'*: John Clare, *By Himself*, p. 36.

113   *'the poor planters . . .'*: Elizabeth Donnan, *Documents Illustrative of the Slave Trade to America*, Vol. 1 (Carnegie Institution of Washington, 1930), p. 92.

113   *'so great a glut . . .'*: ibid., p. 92.

115   *'1630 acres where I now live . . .'*: Peter Wilson Coldham, *American Wills Proved in London, 1611–1775* (Genealogical Publishing, 1992), pp. 166–67.

116 '*The impulse to mercantile accumulation*': Robin Blackburn, *The Overthrow of Colonial Slavery, 1776–1848* (Verso, 1988), p. 7.

116 '*For long periods of time . . .*': W. G. Sebald, *Rings of Saturn*, p. 194.

117 '*a wide walk . . .*': Harriott Horry Ravenel, *Eliza Pinckney* (Charles Scribner, 1896), p. 54

117 '*My letter will be . . .*': ibid., p. 54.

120 '*Now the reasons for plantations . . .*': Captain John Smith, *Advertisements for the Unexperienced Planters of New-England, Or Anywhere, Or, The Pathway to Erect a Plantation* (W. Veazie, 1865 [1631]), p. 22.

120 '*the people of Christendom*': ibid., p. 21.

120 '*the natives of those Countries*': ibid., p. 21.

121 '*new World*': John Milton, *Paradise Lost*, p. 103.

121 '*th' American so girt. . .*': ibid., p. 248.

121 '*wilde / Among the trees*': ibid., p. 248.

121 '*th' addition of his Empire*': ibid., p. 191.

122 '*great adventurer*': ibid., p. 264.

122 '*plant*': ibid., p. 31.

125 '*Thus the wealth . . .*': *Sotheby's Shrubland Hall* (Sotheby's & Co., 2006), p. 12.

126 '*I think I had better . . .*': Jane Austen, *Mansfield Park* (Oxford University Press, 1970 [1814]), p. 47.

126 '*modern dress*': ibid., p. 50.

127 '*dead silence*': ibid., p. 178.

128 '*his property and his negroes*': HA93/M3/20, Middleton Archive, Suffolk Archive.

128 '*104 workers . . .*': HA93/M3/23, Middleton Archive, Suffolk Archive.

130 '*The sheer number of objects . . .*': *Sotheby's Shrubland Hall*, p. 17.

130 '*far more lavish . . .*': James Bentley and Nikolaus Pevsner, *The Buildings of England, Suffolk: East* (Yale University Press, 2015), p. 489.

136 '*prison-house*': Fanny Kemble, *Journal of a Residence on a Georgian Plantation* (Bandanna Books, 2015 [1863]), p. 38.

138 '*the first ample, lucid . . .*': Mildred E. Lombard, 'Contemporary

Opinions of Mrs Kemble's Journal of a Residence on a Georgia Plantation', *The Georgia Historical Quarterly*, Vol. 14, No. 4, December 1930, pp. 335–343.

138 '*The particulars into which . . .*': ibid., pp. 335–343.

139 '*We laughed, an' laughed, an' laughed*': Sarah Hopkins Bradford, *Harriet Tubman: The Moses of Her People* (Corinth Books, 1961), p. 53.

V: GARDEN STATE

147 '*he hurried on . . .*': Thomas Frost, *Forty Years' Recollections: Literary and Political* (Sampson Low, Marston, Searle and Rivington, 1880), p. 70.

148 '*We are opposed . . .*': Barbara Taylor, *Eve and the New Jerusalem* (Virago, 1983), p. 180.

149 '*a highly cerebral . . .*': ibid., p. 175.

149 '*the most gorgeous conceptions . . .*': ibid., p. 176.

155 '*where nothing is wasted . . .*': William Morris, *Selected Writings* (G.D.H. Cole, 1948), p. 68.

158 '*common treasury*': Gerrard Winstanley, ed. Christopher Hill, *The Law of Freedom and Other Writings* (Penguin, 1973), p. 77.

158 '*in trance and out of trance*': ibid., p. 89.

159 '*We are urged to go . . .*': Andrew Hopton, ed., *Digger Tracts* (Aporia, 1989), pp. 31–2.

163 '*should look both orderly and rich . . .*': William Morris, *The Collected Works of William Morris*, Vol. 22 (Cambridge University Press, 2012), p. 91.

164 '*an aberration of the human mind*': ibid., p. 90.

165 '*real*': William Morris, ed. Norman Kelvin, *The Collected Letters of William Morris*, Vol. I (Princeton University Press, 1984), p. 459.

166 '*Chaynee oysters*': William Morris, ed. Norman Kelvin, *The Collected Letters of William Morris*, Vol. II, Part B (Princeton University Press, 1987), p. 572.

166 '*were rolling over one another*': William Morris, *Selected Writings*, p. 188.

170 'the radical principles . . .': Fiona MacCarthy, *William Morris: A Life for Our Times* (Faber, 1994), p. 169.

170 'Well, what I mean by Socialism . . .': William Morris, *Selected Writings*, p. 655.

171 'how can we bear . . .': William Morris, *The Collected Works of William* Morris, Vol. 22 (Cambridge University Press, 2012), p. 49.

172 'sordid, aimless, ugly confusion . . .': William Morris, *Selected Writings*, p. 658.

172 'thriving and unanxious life': ibid., p. 659.

172 'It is the province of art . . .': ibid., p. 659.

175 'Peach and Pear in ruddy lustre glow': John Clare, ed. Eric Robinson and David Powell, *The Early Poems of John Clare, 1804–1822*, Vol. 1 (Clarendon Press, 1989), p. 46.

175 'a compleat harbour both for shade or smell': ibid., p. 47.

175 'jocolately dusk': ibid., p. 47.

176 'Poverty has made a sad tool . . .': John Clare, *John Clare By Himself*, p. 162.

177 'In the past times . . .': William Morris, *Selected Writings*, pp. 147–148.

178 'we are but minute links . . .': William Morris, ed. Norman Kelvin, *The Collected Letters of William Morris*, Vol. II, Part A, p. 248.

180 'increasing distaste': E. P. Thompson, *William Morris: Romantic to Revolutionary* (Merlin Press, 1977), p. 249.

VI: BENTON APOLLO

191 'we felt he would approve . . .': Mark Rumary, in Jill Norris & Ann Banister, eds., *How Does Your Garden Grow?* (Wolsey Press, 2000), p. 11.

192 'As the day emerged . . .': Stella Gibbons, *Cold Comfort Farm* (Penguin, 1938), pp. 213–215.

197 'Rosa mundi, rose of the world . . .': Derek Jarman, *Modern Nature*, p. 4.

197 'I let it run all over . . .': ibid., p. 9.

198 *'the codes and counter-codes . . .'*: *Jubilee*, dir. Derek Jarman (1978)

199 *'curriculum vitae Art History self portrait romance'*: Tony Peake, *Derek Jarman* (Little, Brown, 1999), p. 17.

199 *'bedraggled little boys in formless grey suits'*: Derek Jarman, *Modern Nature*, p. 22.

200 *'blotched'*: ibid., p. 28.

200 *'happiest with the axe'*: ibid., p. 45.

201 *'the park and idyllic acres . . .'*: ibid., p. 57.

202 *'such was our happy garden state'*: ibid., p. 38.

202 *'like two dogs'*: Derek Jarman, *At Your Own Risk* (Vintage, 1993), p. 20.

202 *'like an empty shell . . .'*: Derek Jarman, *Dancing Ledge*, p. 63.

202 *'All the Cains and Abels . . .'*: Derek Jarman, *Modern Nature*, p. 84.

203 *'before I finish I intend . . .'*: ibid., p. 23.

203 *'all of them looking backward . . .'*: ibid., p. 25.

204 *'curious'*: Derek Jarman, *derek jarman's garden* (Thames & Hudson, 1995), p. 12.

205 *'a therapy and a pharmacopoeia'*: ibid., p. 12.

205 *'a fortress against another reality . . .'*: Derek Jarman, *Modern Nature*, p. 61.

210 *'exceeding sweet of smell . . .'*: John Gerard, *The Herball*, p. 51.

210 *'This Iris hath his flower . . .'*: ibid., p. 51.

210 *'notable not only . . .'*: Nicholas Moore, *The Tall Bearded Iris* (W. H. & L. Collingridge, 1956), p. 79.

211 *'Lett and Cedric were open . . .'*: Gwyneth Reynolds and Diana Grace, eds., *Benton End Remembered: Cedric Morris, Arthur Lett-Haines and the East Anglian School of Painting and Drawing* (Unicorn Press, 2002), p. 151.

212 *'the prettiest ones . . .'*: Derek Jarman, *Modern Nature*, p. 114.

212 *'the skies over Chelsea . . .'*: 'Lord Montagu of Beaulieu', *The Times*, 1 September 2015.

212 *'homosexuals in general . . .'*: Hansard, Vol. 521, 3 December 1953.

213 *'In one way or another . . .'*: Ronald Blythe, *The Time by the Sea: Aldeburgh 1955-1958* (Faber, 2013), p. 203.

213 'almost like stepping into a novel': ibid., p. 4.

213 'The buggers don't know . . .': ibid., p. 130.

213 'Chinese fare and coffee-chocolate . . .': ibid., p. 36.

214 'a bewildering, mind-stretching . . .': Beth Chatto, 'Sir Cedric Morris, Artist-Gardener', Hortus No. 1 (Spring 1987), p. 15.

216 'It is not only the intensity . . .': Ronald Blythe, The Time by the Sea, p. 209.

221 'You were ambitious . . .': Beautiful Flowers and How To Grow Them, dir. Sarah Wood (2021).

222 'I would love to see . . .': Derek Jarman, Modern Nature, p. 310.

VII: THE WORLD MY WILDERNESS

229 'more pleas'd my sense . . .': John Milton, Paradise Lost, p. 233.

232 'a wilderness of little streets . . .': Rose Macaulay, The World My Wilderness (Collins, 1950), p. 53.

235 'the red campion . . .': ibid., p. 253.

236 'will make some use . . .': Eliot Hodgkin, Imperial War Museum, Second World War Artists Archive, File Number: GP/55/407

236 'It has occurred to me . . .': ibid.

238 'Golden lads . . .': William Shakespeare, Cymbeline, Act 4, Scene 2.

242 'I own the world . . .': Abu Waad, The Last Gardener of Aleppo, Channel 4 News (2016).

242 'Flowers help the world . . .': ibid.

244 'pale and inhuman': Iris Origo, Images and Shadows (Pushkin, 2017 [1970]), p. 272.

244 'There will be a great deal . . .': Caroline Moorehead, Iris Origo: Marchesa of Val d'Orcia (Allison & Busby, 2004), p. 115.

245 'a profit-sharing contract . . .': Iris Origo, Images and Shadows, p. 292.

246 'the bad aspect . . .': ibid., p. 295.

246 'Almost everyone in the village . . .': Julia Blackburn, Thin Paths: Journeys In and Around an Italian Mountain Village (Vintage, 2012), p. 70.

246 '*We are nothing and we own nothing*': ibid., p. 71.
248 '*our earthly Paradise*': Iris Origo, *Images and* Shadows, p. 83.
249 '*entirely absorbed . . .*': Caroline Moorehead, *Iris Origo*, p. 195.
250 '*I can't pretend . . .*': ibid., p. 208.
251 '*seven very small . . .*': Iris Origo, *War in Val d'Orcia* (Jonathan Cape, 1951), p. 19.
252 '*A weight has been lifted . . .*': ibid., p. 49.
252 '*tramping up the garden path*': ibid., p. 57.
255 '*As I go from one familiar . . .*': ibid., p. 120.
255 '*very unwise*': ibid., p. 122.
256 '*Life is returning to the medieval . . .*': ibid., p. 127.
256 '*Find on my breakfast tray . . .*': ibid., p. 164.
258 '*It is a strange sight . . .*': ibid., p. 215.
259 '*a party in a dream*': ibid., p. 246.
260 '*How strange it is . . .*': Caroline Moorehead, *Iris Origo: Marchesa of Val d'Orcia*, p. 332.
262 '*patience and endurance . . .*': Iris Origo, *War in Val d'Orcia*, p. 252.
262 '*Resigned and laborious . . .*': ibid., p. 253.
262 '*the new doctrine . . .*': Iris Origo, *Images and Shadows*, p. 338.
263 '*acrimonious*': Benedetta Origo, Morna Livingstone, Laurie Olin and John Dixon Hunt, *La Foce: A Garden and Landscape in Tuscany* (University of Pennsylvania Press, 2001), p. 46.
264 '*sadly . . .*': ibid., p. 47.

VIII: THE EXPELLING ANGEL

273 '*The secretary of state disagrees . . .*': https://infrastructure.planninginspectorate.gov.uk/wp-content/ipc/uploads/projects/EN010012/EN010012-011164-SZC-Decision-Letter.pdf
276 '*taint . . .*': Andrew Marvell, 'The Mower Against Gardens', *The Poems of Andrew Marvell*, p. 133.
276 '*He first enclosed . . .*': ibid., p. 133.
276 '*the mower mown*': ibid., p. 139.
277 '*What wond'rous life . . .*': ibid., p. 157.

278 *'that great and true Amphibium'*: Thomas Browne, ed. Geoffrey Keynes, *The Works of Sir Thomas Brown*, Vol. 1 (Faber, 1928), p. 45.

278 *'Far other worlds'*: Andrew Marvell, *The Poems of Andrew Marvell*, p. 157.

279 *'gay enameld colours mixt'*: John Milton, *Paradise Lost*, p. 95.

283 *'once regarded . . .'*: Fergus Garrett, 'How Great Dixter astounded ecologists', *Gardens Illustrated*, 30 July 2020.

BIBLIOGRAPHY

I: A DOOR IN THE WALL

Gerard, John, *The Herball, or Generall Historie of Plantes* (A. Islip, J. Morton and R. Whitakers, 1633)

Jarman, Derek, *Modern Nature* (Vintage, 1992)

Lees-Milne, Alvilde and Rosemary Verey, eds., *The Englishman's Garden* (Penguin, 1985)

Norris, Jill and Anne Bannister, eds., *How Does Your Garden Grown: An Anthology of Suffolk Gardens* (The Wolsey Press, 2000)

II: PARADISE

Anon., 'Entry of King Charles II into London on his Restoration, May 29, 1660', *The European Magazine and London Review*, Vol. 37 (The Philological Society, January 1800)

Browne, Thomas, *Hydriotaphia and The Garden of Cyrus* (Macmillan, 1929 [1865])

Campbell, Gordon and Thomas N. Corns, *John Milton: Life, Work and Thought* (Oxford University Press, 2008)

Delumeau, Jean, *History of Paradise: The Garden of Eden in Myth and Tradition* (University of Illinois Press, 2000)

Ellwood, Thomas, *The History of the Life of Thomas Ellwood* (J. Phillips, 1791)

Empson, William, *Milton's God* (Chatto & Windus, 1961)

Hill, Christopher, *The World Turned Upside Down: Radical Ideas During the English Revolution* (Penguin, 1975)

—, *The English Revolution 1640: An Essay* (Lawrence and Wishart, 1979 [1940])

—, *Milton and the English Revolution* (Verso, 2020 [1977])

Knott, John Ray, 'Milton's Wild Garden', *Studies in Philology*, Vol. 102, No. 1, Winter 2005, pp. 66-82

Lehrnman, Jonas, *Earthly Paradise: Garden and Courtyard in Islam* (Thames & Hudson, 1980)

Leslie, Michael, ed., *A Cultural History of Gardens in the Medieval Age* (Bloomsbury Academic, 2016)

Macaulay, Rose, *Milton* (Gerald Duckworth & Co., 1957 [1934])

McColley, Diane, *A Gust of Paradise: Milton's Eden and the Visual Arts* (University of Illinois Press, 1993)

Marvell, Andrew, ed. Nigel Smith, *The Poems of Andrew Marvell* (Pearson, 2003)

Michel, Albin, *Eden: Le jardin medieval a travers l'enluminure XIII– XVI siècle* (Bibliothèque nationale de France, 2001)

Milton, John, ed. Barbara K. Lewalski, *Paradise Lost* (Blackwell, 2007)

Moshenska, Joe, *Making Darkness Light: The Lives and Times of John Milton* (Basic Books, 2021)

Moynihan, Elizabeth, *Paradise as a Garden In Persia and Mughal India* (Scolar Press, 1980)

Parkinson, John, *Paradisi in Sole, Paradisus Terrestris* (Metheun, 1900 [1629])

Pepys, Samuel, ed. by R. C. Latham & W. Matthews, *The Diary of Samuel Pepys: Volume One*, 1660 (Bell & Hyman, 1970)

Schulz, Max F., *Paradise Preserved: Recreations of Eden in Eighteenth and Nineteenth Century England* (Cambridge University Press, 1985)

Tigner, Amy, *Literature and the Renaissance Garden from Elizabeth I to Charles II: England's Paradise* (Thames Scolar Pres, 1980)

Walpole, Horace, *On Modern Gardening* (Pallas Athene, 2004 [1771])

Waugh, Evelyn, *Brideshead Revisited* (Penguin, 1962)
Woods, May, *Visions of Arcadia* (Aurum, 1996)

III: A LANDSCAPE WITHOUT PEOPLE

Athill, Diana, *Alive, Alive, Oh! And Other Things That Matter*
(Granta, 2015)
Barrell, John, *The Idea of Landscape and the Sense of Place, 1730–1840:*
An Approach to the Poetry of John Clare (Cambridge University
Press, 1972)
Bate, Jonathan, *John Clare: A Biography* (Picador, 2005 [2003])
Brown, David and Tom Williamson, *Lancelot Brown and the Capability*
Men (Reaktion Books, 2016)
Clare, John, ed. Margaret Grainger, *The Natural History Prose*
Writings of John Clare (Oxford University Press, 1983)
—, ed. Mark Storey, *The Letters of John Clare* (Oxford University
Press, 1986)
—, ed. Eric Robinson and David Powell, *The Early Poems of John*
Clare (Oxford University Press, 1989)
—, ed. Eric Robinson and David Powell, *John Clare: By Himself*
(Carcanet Press, 2002)
Darby, H. C. ed., *A New Historical Geography of England* (Cambridge
University Press, 1973)
Ewart Evans, George, *The Farm and the Village* (Faber & Faber,
1969)
Floud, Roderick, *An Economic History of the English Garden* (Allen
Lane, 2019)
Hartog, Dirk van, 'Sinuous Rhythms and Serpentine Lines: Milton,
the Baroque, and the English Landscape Garden Revisited',
Milton Studies, Vol. 48, 2007, pp. 86–105
Hoskins, W. G., *The Making of the English Landscape* (Penguin, 1971
[1955])
Hunt, John Dixon and Peter Willis, eds., *The Genius of the Place:*
The English Landscape Garden 1620–1820 (Paul Elek, 1975)

Mahood, M. M., *A John Clare Flora* (Trent Editions, 2016)

Mayer, Laura, *Capability Brown and the English Landscape Garden* (Shire Books, 2011)

Richardson, Tim, *The Arcadian Friends: Inventing the English Landscape Garden* (Transworld, 2007)

Sebald, W. G., trans. Michael Hulse, *The Rings of Saturn* (The Harvill Press, 1998)

Symes, Michael, *The English Landscape Garden: A Survey* (Historic England, 2019)

Thompson, E. P., *The Making of the English Working Class* (Penguin, 1968)

Thornton, R. K. R., 'The Flowers and the Book: the Gardens of John Clare', *John Clare Society Journal* No. 1, July 1982, pp. 31–45

Walpole, Horace, *On Modern Gardening* (Pallas Athene, 2004)

Ward, Colin and David Crouch, *The Allotment: Its Landscape and Culture* (Faber, 1988)

Willis, Peter, ed., *Furor Hortensis: Essays on the history of the English Landscape Garden in memory of H.F. Clark* (Elysium Press, 1974)

Wilson, Richard and Alan Mackley, *Creating Paradise: The Building of the English Country House 1660–1880* (Hambledon and London, 2000)

IV: THE SOVRAN PLANTER

Allen, David, 'Daniel Browninge of Crowfield: A Little-Known High Sheriff of Suffolk and the Stowmarket Assizes of 1695', *Suffolk Archaeology and History*, Vol. 39, Part 1, 1997

Austen, Jane, *Mansfield Park* (Oxford University Press, 1970 [1814])

Bell, Malcolm, *Major Butler's Legacy: Five Generations of a Slaveholding Family* (University of Georgia Press, 1989)

Bentham, William, *The Baronetage of England, or the History of the English Baronets, and Such Baronets of Scotland, as are of English Families* (Miller, 1805)

Bettley, James and Nikolaus Pevsner, *The Buildings of England, Suffolk: East* (Yale University Press, 2015)

Blackburn, Robin, *The Overthrow of Colonial Slavery, 1776–1848* (Verso, 1988)

Bushman, Richard L., *The Refinement of America: Persons, Houses, Cities* (Alfred A. Knopf, 1992)

Chorlton, Thomas Patrick, *The First American Republic, 1774–1789* (AuthorHouse, 2012)

Coclanis, Peter A., *The Shadow of a Dream: Economic Life and Death in the South Carolina Low Country, 1670–1920* (Oxford University Press, 1989)

Coldham, Peter Wilson, compiler, *American Wills Proved in London, 1611–1775* (Genealogical Publishing, 1992)

Deitz, Paula, *Of Gardens* (University of Pennsylvania Press, 2010)

Donnan, Elizabeth, *Documents Illustrative of the History of the Slave Trade to America, Vol. 1* (Carnegie Institute of Washington, 1930)

Doyle, Barbara, Mary Edna Sullivan and Tracey Todd, eds., *Beyond the Fields: Slavery at Middleton Place* (University of South Carolina Press, 2009)

Dusinberre, William, *Dem Dark Days: Slavery in the American Rice Swamps* (Oxford University Press, 1996)

Evans, J. Martin, *Milton's Imperial Epic: Paradise Lost and the Discourse of Colonialism* (Cornell University Press, 1996)

Greene, Jack P., Rosemary Brana-Shute and Randy J. Sparks, eds., *Money, Trade and Power: The Evolution of Colonial South Carolina's Plantation Society* (University of South Carolina Press, 2001)

Harper-Bill, Christopher, Carole Rawcliffe and Richard G. Wilson, eds., *East Anglia's History: Studies in Honour of Norman Scarfe* (Boydell Press, 2002)

Hopton, Andrew, ed., *Digger Tracts* (Aporia, 1989)

Kemble, Fanny, *Journal of a Residence on a Georgian Plantation* (Bandanna Books, 2015 [1863])

Lombard, Mildred E., 'Contemporary Opinions of Mrs Kemble's Journal of a Residence on a Georgia Plantation', *The Georgia Historical Quarterly*, Vol. 14, No. 4, December 1930, pp. 335-343

Mintz, Sidney W., *Sweetness and Power: The Place of Sugar in Modern History* (Viking, 1985)

Sotheby's, *Shrubland Hall* (Sotheby's & Co., 2006)

Stoney, Samuel Galliard, *Plantations of the Carolina Low Country* (Dover Publications, 1989 [1938])

Trinkley, Michael, Natalie Adams and Debi Hacker, 'Landscape and Garden Archaeology at Crowfield Plantation: A Preliminary Investigation', Chicora Foundation, Research Series 32, June 1992

Williamson, Tom, *Humphry Repton: Landscape Design in an Age of Revolution* (Reaktion Books, 2020)

V: GARDEN STATE

Baker, Derek W., *The Flowers of William Morris* (Barn Elms, 1996)

Brailsford, H. N., ed. Christopher Hill, *The Levellers and the English Revolution* (Spokesman, 1976)

Frost, Thomas, *Forty Years' Recollections: Literary and Political* (Sampson Low, Marston, Searle and Rivington, 1880)

Hardy, Dennis, *Alternative Communities in Nineteenth Century England* (Longman, 1979)

Hunt, Tristram, *The English Civil War at First Hand* (Penguin, 2011 [2002])

MacCarthy, Fiona, *William Morris: A Life for Our Times* (Faber, 1994)

Marsh, Jan, *William Morris & Red House* (National Trust Books, 2005)

Morris, William, *Selected Writings* (G.D.H. Cole, 1948)

—, *The Collected Letters of William Morris*, ed. Norman Kelvin, Vols. I–IV (Princeton University Press, 1984–1996)

Morton, A. L., *The English Utopia* (Lawrence and Wishart, 1969)

Rees, John, *The Leveller Revolution: Radical Political Organisation in England, 1640–1650* (Verso, 2017)

Rodgers, David, *William Morris at Home* (Ebury Press, 1996)

BIBLIOGRAPHY

Taylor, Barbara, *Eve and the New Jerusalem: Socialism and Feminism in the Nineteenth Century* (Virago, 1983)

Thompson, E. P., *William Morris: Romantic to Revolutionary* (Merlin Press, 1977)

Winstanley, Gerrard, ed. Christopher Hill, *The Law of Freedom and Other Writings* (Pelican, 1973)

VI: BENTON APOLLO

Blythe, Ronald, *At the Yeoman's House* (Enitharmon Press, 2011)

—, *The Time by the Sea: Aldeburgh 1955–58* (Faber, 2013)

Gibbons, Stella, *Cold Comfort Farm* (Penguin, 1938)

Jarman, Derek, *Dancing Ledge* (Quartet Books, 1984)

—, *At Your Own Risk* (Hutchinson, 1992)

—, *Chroma* (Century, 1994)

—, *Smiling in Slow Motion* (Century, 2000)

—, with photographs by Howard Sooley, *derek jarman's garden* (Thames & Hudson, 1995)

Melville, Derek, *Derek Melville's Carols* (Mark Rumary, 2001)

Moore, Nicholas, *The Tall Bearded Iris* (W. H. & L. Collingridge, 1956)

Peake, Tony, *Derek Jarman* (Little, Brown, 1999)

Reynolds, Gwyneth and Diana Grace, eds., *Benton End Remembered: Cedric Morris, Arthur Lett-Haines and the East Anglian School of Painting and Drawing* (Unicorn Press, 2002)

St Clair, Hugh, *A Lesson in Art & Life: The Colourful World of Cedric Morris & Arthur Lett-Haines* (Pimpernel Press, 2019)

VII: THE WORLD MY WILDERNESS

Bardgett, Suzanne, *Wartime London in Paintings* (Imperial War Museum, 2020)

Blackburn, Julia, *Thin Paths: Journeys In and Around an Italian Mountain Village* (Vintage, 2012)

Butler, A. S. G., *Recording Ruin* (Constable, 1942)

Fitter, R. S. R., *London's Natural History* (Collins, 1945)

Gardiner, Juliet, *The Blitz: The British Under Attack* (Harper Press, 2010)

Hobhouse, Penelope, 'The Gardens of the Villa La Foce', *Hortus* Vol. 3, Autumn 1987

Hughes-Hallett, Lucy, *The Pike* (Fourth Estate, 2013)

LeFanu, Sarah, *Rose Macaulay* (Virago, 2003)

Macaulay, Rose, *The World My Wilderness* (Collins, 1950)

Moorehead, Caroline, *Iris Origo: Marchesa of Val d'Orcia* (Allison & Busby, 2004)

Origo, Benedetta, Morna Livingstone, Laurie Olin and John Dixon Hunt, *La Foce: A Garden and Landscape in Tuscany* (University of Pennsylvania Press, 2001)

Origo, Iris, *War in Val d'Orcia* (Jonathan Cape, 1951)

—, *Images and Shadows: Part of a Life* (Pushkin, 2017 [1970])

Patterson, Ian, *Guernica and Total War* (Profile, 2007)

Richards, J. M., ed., *The Bombed Buildings of Britain* (The Architectural Press, 1947)

Ward, Lawrence, *The London County Council Bomb Damage Maps 1939–1945* (Thames & Hudson, 2015)

VIII: THE EXPELLING ANGEL

Atkins, William, 'On Sizewell C', *Granta 159: What Did You See?*, Spring 2022

Lloyd, Christopher, *Succession Planting for Adventurous Gardeners* (BBC Books, 2005)

Marvell, Andrew, *The Poems of Andrew Marvell* (Pearson, 2003)

Rumary, Mark, *The Dry Garden* (Conran Octopus, 1994)

ACKNOWLEDGEMENTS

There are thanks due for the book and thanks due for the garden. To begin with the two entwined: what good fortune to find myself living next door to John Craig, who made the beautiful illustrations of the garden through the seasons, and who has been such a friend to me, and to the late Pauline Craig, who loved flowers even more than I do. To Matt Tanton-Brown, who has done so much to transform the garden. And to Lynn and John Walford, who protected it so well.

In England: Rebecca Carter, agent and friend. To the team at Picador, particularly my brilliant editor Mary Mount, the endlessly reassuring and now sadly departed Gillian Fitzgerald-Kelly, Ebruba Abel-Unokan, for much great fielding, Stuart Wilson, whose cover design knocked it out the park, Lindsay Nash, for her magical text design, Elle Gibbons, Eloise Wood, Laura Marlow, Connor Hutchinson, Emma Bravo, who went well beyond the call of duty, and the meticulous Nicholas Blake. To all at Janklow, especially Kirsty Gordon and Mairi Friesen-Escandell. And to the team at Strathmore Studios and especially Cherry Cookson, for making recording the audiobook such a pleasure.

In America: PJ Mark, also agent, also friend. To all at Norton, particularly Jill Bialosky, who asked such piercing questions, Drew Weitman and Laura Mucha, Sarahmay Wilkinson and Kelly Winton, who came up with such an exuberant cover, Erin Sinesky-Lovett, gardener-publicist, and Steve Coalca for all his work.

Major thanks to Sam Talbot and his team, especially Kitty Malton and Alicia Lethbridge.

To the people who answered my queries, provided research assistance, or helped me get into the gardens I longed to see. Thank you to the staff at the British Library, the Imperial War Museum, the London Library (especially Claire Berliner), Lowestoft Archives, Royal Horticultural Society, Suffolk Archives, Tate Library and Archive and the Warburg Library. To Christopher Woodward and Matt Collins at the Garden Museum for showing me Benton End, and to Ollie Whitehead for creating a Beltane celebration. To Fergus Garrett and all at Great Dixter, and to Mair Bosworth for bringing me there in the first place. To Amanda Wilkinson, Howard Sooley and the late Keith Collins, for Prospect Cottage. To Sarah McCrory and Michael Cioffi, for introducing me to La Foce. To Joost Depuydt and the Museum Plantin-Moretus, for so generously providing some of the same woodcuts that appear in Gerard's *Herball*. To Simon Bagnall, at Worcester College, Oxford. To Nick Cullinan and Mary Beard for fortuitously directing me to the current location of the Villa Livia garden. And for answering questions about Mark Rumary: Lady Caroline Blois, Rupert Eley, Roger Gladwell and his troops, Diana

Howard and John Morley, Carole Lee, Lee Mayhew, Peter Manthorpe, John and Amanda Sutherell, Richard and Rita Walker: thank you.

To the National Garden Scheme and their stalwart work, and especially Jenny Reeve and Michael Coles. And to the open garden team: Lorraine and Margaret Strowger, Rebecca May Johnson and Sam Johnson-Schlee.

A special thank you to the friends who've talked with me about gardens and paradises, who've discussed ideas, made suggestions, provided sustaining distractions and read early drafts. Charlie Porter, words can't do justice to how grateful I am for you. Richard Porter, thank you for that beautiful pilot light. Chantal Joffe, proud to be shining my shoes with you. Francesca Segal, sustenance & the spreadsheet, seventh time round. Sarah Wood, you make me more ambitious for the imagination! And to Julia Blackburn, Nick Davies, Brian Dillon, Jean Hannah Edelstein, Tom de Grunwald, Philip Hoare, Lauren John Joseph, Emily LaBarge, Lili Stevens, Carole Villiers, Matt Wolf: so much love and gratitude to you all.

The seed of this book first emerged as an essay for the *Observer*. Dan Franklin asked me to write about the garden for the anthology *A Suffolk Garland*, Amy Sherlock about Benton End for *World of Interiors*, and Thea Lenarduzzi about my own utopian past for the *Times Literary Supplement*. Thank you all.

Thanks too to my family: Denise Laing, Kitty Laing, Tricia Murphy and especially my father, Peter Laing, who passed on his passion for gardens and as such appears in this book rather more than he might have liked.

And of course, to Ian Patterson, the librarian, who was there through it all. It's a *boke* and an *appill*, for you.

LIST OF ILLUSTRATIONS

Linocuts by John Craig, 2022–23

Spring
Winter
Summer
Autumn

Woodcuts from the Collectie Stad Antwerpen, Museum Plantin-Moretus, 1568

Flower basket, MPM.HB.09066
Paradise apple, *Malus pumilla*, MPM.HB.04803
Wallflower, *Erysimum cheiyri*, MPM.HB.06303
Sugar cane, *Saccharum officinarum*, MPM.HB.04896
Florentine iris, *Iris florentina*, MPM.HB.07910
Rosebay willowherb, *Chamerion angustifolium*, MPM.HB.05622
Wild thyme, *Thymus serpyllum*, MPM.HB.05019

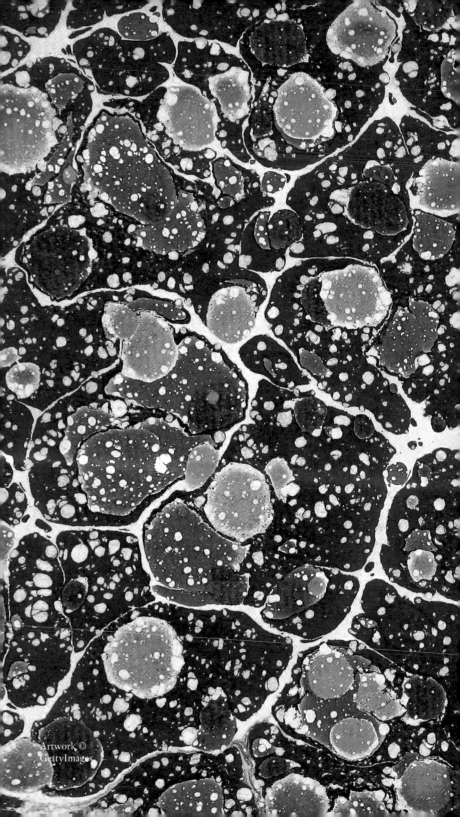